A Kind of Courage

COLLEEN HEFFERNAN

ORCA BOOK PUBLISHERS

National Library of Canada Cataloguing in Publication Data

Heffernan, Colleen, 1959-
A kind of courage / Colleen Heffernan.

ISBN 1-55143-358-3

I. Title.

PS8565.E325K55 2005   jC813'.6   **C2005-903590-0**

First published in the United States 2005
Library of Congress Control Number: 2005928637

Summary: When a young conscientious objector comes to work on her
father's farm in 1916, Hattie learns that courage comes in many forms.

Orca Book Publishers gratefully acknowledges the support for its publishing
programs provided by the following agencies: the Government of Canada
through the Book Pulishing Industry Development Program (BPIDP),
the Canada Council for the Arts, and the British Columbia Arts Council.

Cover design and typesetting: Lynn O'Rourke

Cover artwork: *image 1*: Glenbow Archives NC-39-156;
image 2: Glenbow Archives bapost 2012;
image 3: Private Collection;
image 4: (backcover) University of Toronto Libraries, JB Tyrrell Collections Fo345

In Canada:	**In the United States:**
Orca Book Publishers	Orca Book Publishers
www.orcabook.com	*www.orcabook.com*
Box 5626, Stn. B	PO Box 468
Victoria, BC Canada	Custer, WA USA
V8R 6S4	98240-0468

08 07 06 05 • 5 4 3 2 1
Printed and bound in Canada
Printed on 100% post-consumer recycled paper,
100% old growth forest free, processed chlorine free using vegetable, low VOC inks.

For J. Dale Armstrong,
brilliant and generous educator,
my model for integrity, for excellence and for courage.

Acknowledgments

There are many people to whom I owe a debt of gratitude for the background stories that made this book possible—individuals who told me personal stories, others who wrote letters, diaries and reminiscences, and professional historians whose thoughtful work gave me a better understanding of this important time in our history. I wish to extend thanks to all of them for enriching my life and this story.

In particular, I wish to thank Barb Neil for sharing with me the diaries and letters of her great uncle, Russel Scobie McAllister; Ken Tingley for his edition of the letters of Alwyn Bramley-Moore and for generous professional assistance with historical research; to the late Dr. Barbara Roberts, whose impassioned biography of Gertrude Richardson gave me Hattie; and most especially to my beloved mother-in-law, Madeline Fielding, for telling me the stories that started it all.

"The causes of war are always falsely represented; its honour is dishonest and its glory meretricious, but the challenge to spiritual endurance, the intense sharpening of all the senses, the vitalizing consciousness of common peril for a common end, remain to allure those boys and girls who have just reached that age when love and friendship and adventure call more persistently than at any later time... while it lasts no emotion known to man seems as yet to have quite the compelling power of this enlarged vitality."

— Vera Brittain
Testament of Youth

Prologue, 1910
HATTIE

"HATTIE? ARE YOU in there, Hattie?" Will lay down on his stomach and crawled through the maze of branches they had threaded together to camouflage the entrance to their secret cave. "Gol dang it, Hattie. Why'd you run off with Jimmy's quarterstaff and ruin the game?"

Hattie didn't want to look at her brother, didn't want him to see her moist eyes. "And why'd you take his side? 'Cause he's your friend and I'm just your dumb little sister?"

Will shook his head. "I never said that."

"Jimmy did." Hattie's voice squeaked. "Every time Jimmy gets to play, he's Little John, 'stead of me—and Herbie's Allen a Dale and Frankie's Will Scarlett and Tom's Friar Tuck.

"First, they don't want me to play. Then, if I play, they only want me to be Maid Marion and sit in the dungeon waiting to be rescued or back at the tree house cooking venison stew.

"I don't want to be Maid Marion. I want to be Little John. I can knock the blocks off the Sheriff's men as good as Jimmy."

Will shook his head again. "I know you can. But you gotta understand—a fellow can't let his sister be Little John."

"You can too. You just don't want to." Hattie gripped the quarterstaff fiercely and turned it in her hands.

Palms up, he beckoned to his sister. "Come on, Hattie. The lads would laugh at me."

Hattie took the quarterstaff firmly in both hands and threw it to him. "Take what you came for and get back to your friends."

Will winced. "You and me will always be the best of friends, the very best. I swear it—like we always have."

Hattie pretended not to hear. Will backed out of the cave on his hands and knees, dragging the quarterstaff in front of him. When Hattie couldn't see him anymore, she leaned forward and yelled at the brambles.

"Don't you say it, Will Tamblyn, 'cause it ain't true and you know it. It ain't true." She sat back on her haunches and whispered, feeling the drops slide down her cheeks. "Not anymore."

April 1918
HATTIE

HATTIE JUMPED TO HER feet and grabbed her blue sweater as soon as she saw Dada drive the wagon into the yard. It was her turn for a letter from Will and she was sure there would be an envelope for her with the YMCA mark in the left corner. In her excitement, she almost slid down the banister like she and Will used to do when they were kids.

The thought of how her mother would react stopped her in time. Everything seemed to pain Mima these days. Her hair hadn't even been gray in 1916 when Will and Jimmy and Frankie and all their friends enlisted. Now it was almost white.

Hattie stopped short at the kitchen door. Her mother stirred a large kettle on the stove. Her father stood close beside her and they talked in hushed tones.

Dada looked over Mima's shoulder. "Morning."

"Was there mail for me?"

"No." Dada shook his head. He pushed up his shirtsleeves and poured water from the pitcher into the basin. Laying his hands flat in the water, he grabbed the bar of soap and rubbed it carefully between his hands. Hattie had watched him and counted ever since she was a little girl.

He was always precise—rubbed the soap between his hands sixteen times, then squirted it from between his hands into the dish, rubbed one palm over the back of the other hand, then switched hands and did it again. He cleaned each nail carefully with the file, starting with the pinky on the left hand and finishing with the thumb on the right. Hattie must have watched him do this a thousand times and he never once varied his method.

"Hattie," he said, patting his hands dry with a towel, "I want you to fix up the room in the barn loft for a hired man. He's coming this afternoon."

"Scrub it up good," her mother added. "Clean bedding and towels. And you can put Aunt Delphine's basin and pitcher up there. It's in the cold room beside the jars of applesauce."

"Where'd you find him?" Hattie asked. Her father had been looking for help on the farm almost since Will left for France. Her younger brother Johnny was thirteen and big as a man, but Dada was fierce that he stay in school. Men willing to work on farms were scarce. Dada had tried a few, but none had worked out. Some were just biding their time until they could sign up to go overseas, some were just too sickly to do the work. He let the last one go at harvest time after he disappeared on Saturday night and Dada and Johnny found him Sunday morning passed out from drink in the gutter in back of the livery stable in town.

Now, her father coughed and rubbed his chin as an answer to her question.

"Sergeant Murphy knew we needed help, what with planting to do, so he's put us in touch," her mother said.

Paddy Murphy was an old friend of her father's from Havelock. He'd joined the army and was now at the local depot. So it made sense that he'd help out, but Hattie couldn't help but feel that something was out of place, that something wasn't quite right.

Hattie took off her blue sweater; she didn't want to get it filthy sweeping out the barn loft. Trading it for one of her mother's ample smocks, she took the broom, a mop, some rags and a scrub pail half full of hot water and headed across the yard.

Her father caught up with her out of sight of the kitchen window. "Wait," he said, pulling a card from the inside pocket of his coat. "This was in the post for you. But don't tell your mother—you know how worried she gets."

Hattie set down the pail and took the card—a field post-card from Will. She nodded to her father. "Mima was fine before Mrs. Nelson told us soldiers only send field cards just before a battle where they might be killed."

"Well, it's best she don't see it." Dada turned toward the horses still hitched to the wagon and Hattie slipped the card into her pocket. She went into the barn and climbed the makeshift stairs to the loft. Putting everything but the broom in one corner, she leaned on it, thinking about the job ahead of her.

The loft was dirty after a winter of disuse: mostly dust, cobwebs and mouse droppings. As she attacked it with the broom, sweeping everything out the door and off the steps to the barn floor below, she tried not to look at the walls. The boards, fastened together with too many nails and not quite on the square, always reminded her of Will.

He'd built this room, with Dada's permission, the summer he was fourteen.

She remembered him shinnying up the thick rope suspended from the barn roof and swinging on it until he could launch his body through the door into his room. Hattie had looked up at him enviously until he shouted. "Hattie, come on, you can do it."

Her mother would have pitched a fit at the sight of her daughter, skirts hitched up, swinging from a rope like a hooligan. But Mima rarely went into the barn, so Hattie had been free to follow her brother into the loft.

The stairs had come later, built by her father after Will had sailed for England and Mima refused to have a hired man sleep in the house.

Every swish of the broom brought back memories she tried to avoid: planning adventures, reading secret books, avoiding extra chores and their little brother. By the time she finished sweeping and scrubbing the place into shape, her heart was like a path with too many footprints on it.

She pulled the postcard out of her pocket and studied it for clues. She pressed it between her hands trying to imagine where he was when he wrote it and what he would have said if he'd been able to write a letter. Maybe it was the long distance the card had traveled or the many people who'd handled it, but Hattie could find no trace of her brother on the card. Even his signature and the date at the bottom in his own hand seemed rushed and formal.

**NOTHING is to be written on this side except
the date and signature of the sender. Sentences
not required may be erased. If anything else is
added the post card will be destroyed.**

[Postage must be prepaid on any letter or post card
addressed to the sender of this card.]

I am quite well.

I have been admitted into hospital

{ *sick* } *and am going on well.*
{ *wounded* } *and hope to be discharged soon.*

I am being sent down to the base.

I have received your { *letter dated* Feb 25
{ *telegram* „ _____
{ *parcel* „ _____

Letter follows at first opportunity.

I have received no letter from you

{ *lately.*
{ *for a long time.*

Signature } Pte. William Tamblyn
only }

Date   Mar. 27/18

(94800) Wt. W1566-R1619 14,000m. 6/17. J.J.K. & Co., Ltd.

She hated fixing this room up for some stranger instead
of for her brother. Maybe if she imagined she was arrang-
ing it for Will, it would be easier to do.

With the cleaning supplies lugged back to the house,
Hattie dug in the linen closet for sheets, blankets and a
runner for the top of the dresser. She polished up the basin
and pitcher and found an old flower vase. Maybe she could
find some trilliums.

As she staggered through the kitchen loaded with supplies, her mother added a book to the pile. "Put this away somewhere special, dear. Mr. Wilson dropped it by for Will, asked if we'd send it with our next parcel. Thought he might like it, I suppose. Wasn't that nice of him?"

"Yes, Mima." Whenever her mother thought about Will these days, the rest of the world didn't exist. Couldn't her mother see that she was loaded down with things for the loft and didn't have time to find a special place for Mr. Wilson's book? But it was best not to point that out. "I'll send it as soon as I finish the balaclava and that last pair of socks."

In the loft, Hattie arranged the basin and pitcher and the vase on the dresser. She laid down fresh straw and made up the bed before she took a close look at the book, *A Knight on Wheels,* by Ian Hay.

"Oh, that's funny," she said out loud. Will would have thought it was funny—reading a book by Hay on a straw bed.

THE WIND WAS GENTLE and the sun warm on her back but Hattie could find no trilliums. So she wandered into the woods where she and Will had played Robin Hood when they were kids. It wasn't an accident. She wanted to hide the field card in the stone box in their old hideout.

This cave seemed so big when we were kids, she thought as she knelt, feeling with her hand for the box. We'd haul branches to camouflage the entrance and hide from the Sheriff's men.

There, the smooth touch of soapstone. It was a keen box. Their Uncle Fred, Dada's oldest brother, who'd been everywhere, even to the gold rush in the Yukon, had brought them a block of stone. He showed them how to cut a lid with a hacksaw and to sharpen old spoons with a file to scoop out the inside.

The day Will left for Halifax, Hattie brought the box out here, vowing to keep every letter Will sent her in this charmed place. She put the field card on top of the last letter she'd received, the one she wished she hadn't, the one she knew almost by heart.

Dear Hattie,

By the time you get this letter, you'll know that Frankie Hamilton was killed. Writing to his mother, telling her how brave he was and all, was the hardest thing I ever had to do. But if this gets past the officer and his black pen, I can tell you how it really was and maybe I won't feel like such a godawful liar.

The battalion was in reserve and three of us, Frankie and I and a sapper from Kingston, were in this bombed out house. I had orders not to leave the telegraph and Fritz was pretty quiet, so we played a few rounds of our favorite game.

Frankie'd been winning our money and was feeling pretty chipper when all of a sudden, he rolled snake eyes.

"Dang it to hell," he says, "I'm jinxed now. I'd better lie low today."

We kept playing, but he was pretty jittery. It didn't help when Fritz sent a few coal boxes over on the line.

A few minutes later, a medic comes by. He needs someone to help him carry a stretcher to the aid station. I look at the sapper, who's busy trimming his fingernails with a pocketknife. I can tell he's not going to go. And I'm supposed to stay with the telegraph, which leaves Frankie.

But when I look at Frankie, his eyes are big and round like marbles and his face is almost gray. He's still holding the dice in one hand and they're jiggling, dancing in his hand, but I can tell he's not doing it on purpose.

"Frankie." I bang him on the shoulder. "Frankie, I'm going to help out the medic here, okay? You listen for the telegraph, you hear?"

Frankie snapped out of it a little, gave me a bit of a grin.

"And if the officer comes by," I say, "tell him I went for a whizz."

I grab one end of the stretcher, and we slog it through the mud back to the first aid station. Jerry had upped his shelling and whizzbangs exploded all around us. The medic said for me to wait until it slowed down, but I had to get back to the telegraph.

It only took me a half hour to go to the aid station and back, but when I got there, what was left of the house was gone. Just a pile of crumbled stone. The sapper and Frankie were both dead. I could hardly stand to look at him, like it was my fault for stealing his good luck. But maybe if he'd have taken the stretcher, he'd have got it anyway. I took his diary and stuff from his pocket to send to his ma and pa.

And when the detail came to bury him and they lifted him onto the stretcher, the dice fell from his hand into the mud.

I'm sending them to you 'cause they're sure bad luck over here. Keep them safe until I come home. Give my love to Mima, Dada and Johnny, but don't tell them about Frankie.

Love, Will

Hattie returned the box and its contents to the cave and strolled over the hill, away from the woods. The last time she'd read that letter she'd thought she saw something in the woods. A flash of red in the middle of winter couldn't be easily explained. Except in their games, when they were kids, Frankie had always been Will Scarlett, worn a cape made from an old red petticoat. Hattie shuddered. She didn't want to think about it.

From around the bend in the road where the bridge crossed the Indian River, the army truck full of soldiers slowed, then stopped at the farm gate.

"Will." Hattie thought at first, then realized it was foolish.

A tall broad man barked an order. Two soldiers grabbed a third by the arms and threw him from the truck. He staggered backwards and fell heavily in the mud and gravel of the drive.

Hattie held her breath as he lay there for a few seconds. Then pushing himself up on his elbows, he turned to get up. Hattie, lifting her skirts, ran to find out what was going on.

Last fall's grass was brown and soft under her feet, the pungent smell of decay fresh in her nostrils. Hattie thought she recognized one of the men in the truck, Pete Harker, a boy they'd gone to school with. He lifted a heavy canvas kit bag and hurled it at the man they'd thrown off the truck, knocking him backwards again onto the gravel.

"Hey!" Hattie yelled, lifting her arm as if she could stop what was happening fifty yards away from her.

Pete Harker yelled at the man on the ground. "Now you'll have to carry it, won't ya?"

The truck roared off down the concession. Pete tipped his cap at Hattie as his mates started singing, "It's a long way to Tipperary, It's a long way to go."

By the time Hattie had reached the farm gate, the young man had picked himself up and was brushing the mud from his long khaki pants.

"Are you all right?" Of all the questions she wanted to ask, this was the only one Hattie would allow herself. Who was this young soldier? Why did the others treat him so roughly? And what was he doing here?

He raised a pair of serious brown eyes with a look of directness that made Hattie regret she'd asked that much.

"I'm David Ross. Mr. Tamblyn is expecting me." He raised his right hand as if to offer it to her. A quick glance seemed to make him think better of it and he grabbed his kit bag with it instead. Swinging it over his shoulder, he stood still, as if waiting for Hattie, his grave eyes looking at her.

"Dada's in the barn." She hesitated, twisting the tie of Mima's smock with one hand. "I'll take you to him, if you like." He nodded. "I'm Henrietta," she said, as an afterthought, "but they call me Hattie."

He was already walking up the drive when Hattie saw the wound on the back of his head, the bloody mass almost hidden by his chestnut hair, the edge of his collar dark with it.

Hattie tried to say something, but all she managed was a strangled squeak. What was wrong with her this morning? She was never hesitant, always bold, much too bold for Mima's liking. Something about this tall mysterious young man made her unsure of herself.

His long strides were more than Hattie could match. She lifted her skirts and ran after him up the long steep drive. He was already talking to her father when she reached the barn.

Dada was his usual reserved self.

"Hattie will show you where to put your things," he said with a nod toward the loft room.

"But he's wounded," Hattie blurted out, finally finding her voice. If she had stopped to imagine what their reaction would be to her statement, she would never have guessed correctly. David's face twisted as if in pain and his dark eyes looked positively haunted.

Her father looked at her, Hattie thought, as if she'd committed a mortal sin right there on the barn floor, instead of merely stating the obvious.

"On the back of your head, Mr. Ross," Hattie said.

He lifted his hand and felt the spot gingerly, then inspected his bloodied fingers. "It's just a scratch, Mr. Tamblyn," he said. "I slipped and fell on the drive." He indicated the mud on his uniform as if it was proof of his lie, Hattie thought.

Dada looked at her sadly. "Take him to his room, get him some warm water and salts to clean that up, then bring him down to the pasture. We'll be working on the fence."

LATER THAT AFTERNOON, Hattie's mother sent her for the eggs. At the chicken coop, Johnny sat on the step, a pail of eggs beside him.

"Mima thought you weren't home," she told him. "You'd better get up to the house."

He turned slowly toward her. "I can't."

"Good gracious, what happened?" His right cheek and eye were swollen and pink. A cut above the eyebrow had dripped blood down the side of his face. Hattie put one finger under his chin and lifted to get a better look. "That's going to be a black eye."

"Yeah." He grimaced.

"What's the matter? Usually you're proud as punch when you come home with a shiner."

Johnny made no reply, just hunched over and covered his face with his hands. Hattie couldn't understand it. Johnny was always getting into scraps, but he never worried about it. Like water off a duck's back, Mima always said. Hattie didn't like the fighting, didn't think that Johnny should have to do that. Will never had. All the same, she'd never seen him upset about it.

"Come on," she coaxed, "why don't you tell me how bad the other guy looks."

At first Hattie thought it was a hiccup she heard, then she realized it was a sob. "I'll get some water, fix you up."

When she got back, he let her sponge the blood from his face and neck. He didn't even wince when she put disinfectant on the cut and covered it with a plaster. It was an epidemic of bravery, Hattie thought. David Ross had acted the same when she'd cleaned his gash.

"Who hit you, Johnny?"

His jaw clenched as if to hold back the words.

Judging by the swelling on his face, it was someone with a powerful fist and a venomous rage. Then she knew. "It was Thatch, wasn't it?"

"No!" His voice was louder that it needed to be.

"Stook, then."

He shook his head. She made him look her in the eye.

"Was it both of them?" It would be just like those Hamilton twins to pick on someone younger. Mima said they'd been wild from birth—the story was that Mrs. Hamilton had stopped during harvest only long enough to deliver them before she was back in the fields stooking grain. And Will never had any use for them. One spring after he'd caught them torturing a raccoon in Sherwood Forest, Will had a fight with Frankie over it. Hattie remembered him telling Frankie to "keep those hellion brothers of his off our property."

Johnny looked her straight in the eye. "No."

"You are getting to be a good liar, aren't you? This has Hamilton written all over it. I'm not going to let those bullies get away with this, no matter how big a man you think you are."

The swelling around his eye seemed to throb. "Don't, Hattie. It wasn't Thatch, I swear." His voice dropped to a mere whisper. "It was Billy."

"How can I believe that?" Hattie sponged the blood spots on his shirt. Billy Hamilton, Frankie's youngest brother, was the quietest kid on the line. He was also Johnny's best friend.

"He said Dada's getting a conchie to work on the farm."

"What?" Hattie couldn't quite understand what she'd heard.

"I said it wasn't true, that Dada would never have a damn slacker on the place. Not with Will in France, not with me to help on the farm." Johnny's voice raised in pitch.

Everything was starting to fall into place: the soldiers in the truck, how they'd thrown David Ross into the gravel, how he'd looked when she'd said he was wounded. Oh God, she'd said he was wounded.

"You listening, Hattie?" Johnny looked at her quizzically.

"Yeah, I'm listening. But why did he hit you? Did you hit him first?"

"No!" It would have been a mark of shame for Johnny to hit first when his opponent was smaller and younger than he was. His face fell and his shoulders slumped. He barely whispered. "I said Frankie was a coward, that he was afraid and didn't die like a hero at all."

"Johnny! Why?"

"I don't know," he blurted out. "I knew right away I shouldn't have."

Hattie looked at him, but he turned his eyes away. Then her stomach sank and she knew what he'd done. She took him by the chin and made him look her in the eye. "You read Will's letter, didn't you?" Silence. "Didn't you?"

"Yes."

"Did you tell Billy?"

"Hey, that hurts."

She loosened her grip on his chin. "Just tell me."

"No, I didn't have time. He walloped me with a rock."

"You listen to me, John Patrick Tamblyn. Don't you ever tell about that letter. You understand?"

"Yes."

"I'm going to burn it and that'll be the end of it. The Hamiltons have enough grief."

HATTIE HAD NEVER seen Johnny this miserable, not even when Will left on the train. She took the pail of eggs from him as they reached the house.

"You wait here," she said. "I'll take care of it."

Johnny slumped onto the stairs and cupped his chin in one hand.

The kitchen was full of steam from the pot of boiling potatoes. Hattie put the pail on the counter.

"Johnny'd already got the eggs," she said.

"That so?" Her mother eyed her with suspicion. "Then what took you so long? And why didn't Johnny bring the eggs himself?"

"He got a shiner, Ma, in a fight with Billy Hamilton. I said I'd tell you about it."

Mima bent her head to one side and raised her eyebrows at Hattie. "And how could young Billy give our Johnny a shiner? Did he stand on a stool and wallop him with a two-by-four?"

"Something like that."

Mima's eyes flashed. "The little blister. I'll speak to his mother."

"No, Mima, don't." Hattie rested one hand on her mother's shoulder. "Johnny said something awful to Billy, something he had no call to say... about Frankie." The light in her mother's eyes extinguished. "He wasn't just being mean. Billy teased him, said we were getting a con-chie to help on the farm. Don't punish him, he's miserable enough already."

Her mother turned to the stove, stirred the milky water and speared a potato with a long-handled fork.

"It's true, isn't it?" Hattie whispered.

"What is?" Lifting the heavy pot off the stove and balancing the lid to allow a stream of water, Hattie's mother poured the liquid from the pot into the slop for the hogs.

"Mr. Ross is not just a hired man, is he? He's a conchie!" Hattie spat out the hated word.

Mima wiped the steam from her face with the bottom of her apron. "Your father's health is not too good—he needs another man to do the work. And if that young man is here on the farm at least he's doing something that needs to be done instead of cooling his heels in a jail cell. We need him and I don't want to hear another word."

JOHNNY SAT ON THE porch holding a wet cloth to his eye. Hattie sat beside him. The barn door slammed and their father strode toward them with unusual purpose just as an old buckboard rounded the curve by the woods. Johnny groaned.

"Him!" Hattie felt her hackles rise.

The small wiry driver jumped from the wagon and ran to keep up with Dada as he approached the porch. Hattie had never seen her father so angry—every muscle in his face looked tight and the skin was flushed pink. With what Hattie thought was great determination, her father relaxed his jaw and turned to the smaller man.

"What can we do for you, Eli?"

Eli waved his hands. "Not a thing, Mr. Tamblyn. Eli was hoping he could do something for you. You'll be needing help with the fences?"

Dada shook his head slowly. "We're much obliged to you, Eli, for all your help. But no, we won't be imposing any further—hired a new man today."

Eli lowered his eyes, slumped his shoulders. "Eli was hopin' … "

"Hattie, get some eggs for Eli," her father said.

"Oh no, sir, Mr. Tamblyn. You're too good … "

Hattie ran to get the eggs, the quicker to get Eli Hough off the property. He weaseled something out of Dada every time he came—simpering and whingeing. He probably knew about the conchie already—if Margaret Hamilton knew, everyone on the line knew. But Eli would pretend all the same, if he thought he could get something. And he could always get something from Dada. It wasn't that her father was a soft touch all around—he'd had little patience for the hired hand they'd found drunk in town, barely giving him time to throw his few clothes in a bag.

Mima stood beside her and watched her pack the eggs. "So he's back for more, is he? Worth having a conchie on the place just to be rid of him." Hattie hadn't seen such a blaze in her mother's eyes in months and she took the chance to ask. "Why does Dada give him things all the time? Can't he see that he's a snake?"

The blaze smoldered. "It's her. He still feels sorry for her."

"Eli's mother?" Hattie knew the story—how Mr. Hough, an old bachelor farmer, had gone away one summer and came back with a teenage bride.

"I did too—left a widow a year after her wedding— pregnant and no menfolk to help her out. But that wasn't our job to fix. Your father figured differently." Waving Hattie's hands aside, she tossed three or four eggs on top with just enough force to crack the ones beneath. Laying a fresh cloth on top she handed the basket to Hattie. "There is a limit to Christian charity."

Astonished, Hattie went outside, where Eli was still whingeing.

" … don't need to give Eli something."

"Nonsense," Dada said, passing him the eggs. There was a long moment of silence as the Tamblyns waited for Eli to leave. He never takes a hint, Hattie thought.

"Much obliged to ya, Mr. Tamblyn. Hattie."

Don't slide your greasy smile at me, she thought, as he climbed back in the wagon and clicked to the team.

When he was gone, Dada drilled his eyes into Johnny. "Get me a switch."

Dada had never switched anyone. Not even when Will had rolled off the hay stack and tipped the bucket, breaking all the eggs. He'd been mad and Will did extra chores for a week, but Dada hadn't switched him.

"You heard me—get a switch."

Johnny stood up, fumbled in his pocket for his knife and walked toward the woods. He stopped at the edge and bent to cut a piece of willow.

"Dada," Hattie almost whispered. "He feels real bad already and Billy walloped him good. You don't need to switch him too."

"Stay out of this, Hattie. T'ain't your concern."

That's the trouble, Hattie thought, it *was* her concern. If she'd burnt that letter to start with, none of this would have happened. But if she told now—it'd look bad on Will as well as Johnny. Johnny shambled toward them, dragging the branch. She had to do something.

"Dada, Billy teased him about us getting a conchie, about... Mr. Ross. Johnny didn't know—he thought it was a lie. He thought... " She stopped when she saw the look on Johnny's face.

"You mean it's true," he spat. "A dirty conchie?"

Hattie's father held his hand out for the switch. "You listen to me, both of you. While Mr. Ross is here, helping us on the farm, I don't want to hear the word conchie, nor slacker, nor shirker, nor coward neither. You're to treat Mr. Ross as if he was any other hired hand and that's an end to it. Do you hear?"

Johnny turned on his father with a look of seething hatred. "Mr. Hamilton said he'd let his hay rot on the fields before he'd let a conchie stack it."

"Hugh Hamilton would cut off his nose to spite his face, boy. How do you think Will and all those other boys are going to get fed if we do that? The government may not think we're 'doing our bit,' canceling the farm exemption and all, but they'll be singing a different tune if we don't get the fields planted and harvested."

Calm and gentle by nature, her father did not often get riled up. Now, the exertion seemed to drain the anger from him and Hattie thought she'd try again.

"Johnny didn't know. Don't switch him."

But as her father turned with the switch in his hand, she realized that she'd miscalculated. "I don't care what he knew. And I don't care what disagreements you or I may have with the Hamiltons. There's not a reason in the world to insult their poor dead boy. None."

None, except that what Johnny had said was true. Hattie looked at her brother. His eyes flashed defiance, but he set his lips in a cold straight line. "No, sir, none."

Mr. Tamblyn pointed to the barn with the willow switch and Johnny marched toward it without another word. Hattie sank onto the steps and covered her ears.

A sound reached through, but it wasn't what she'd expected. It was a cow bell—David Ross was leading Nefertiti up the path toward the barn. Lord, it was bad enough Johnny got switched, he didn't need to have the cause of it all bursting in on them.

"Mr. Ross!" She gathered her skirts and ran, planting herself between him and the barn.

"Yes, Miss Hattie?" His look was unsettling. He seemed to know just what she was thinking, although that was impossible.

"Ah … supper." It was the only thing she could think of.

"Supper?"

"Yes, supper will be ready directly." He made as if to continue to the barn, but she stood her ground.

"Your father asked me to bring this one into the barn. We found her down by the river."

Hattie moved to the side but started petting the cow, stroking her neck. "She likes the river. Thinks it's the Nile, don't you? Queen of Egypt, after all."

"Queen of Egypt?"

"Nefertiti, that's her name, silly old cow. Will named her after they found that tomb in the desert. Makes her quite exotic, doesn't it?"

The barn door opened and Mr. Tamblyn strode toward the house, flinging the willow stick away from him as he went. Hattie waited, her hand firmly on the cow's rope, but Johnny didn't come out.

"I'll take her in for you, Mr. Ross. I know where she lives. That way, you can be washed up in time. Mima gets frightful when we're late." That was a lie, but it was all she could come up with on short notice.

He looked puzzled. "Mima?"

"Mum," she explained. "My mother's family were Irish, so we say it like that."

ONCE THE DISHES were done, Hattie wrapped a warm shawl about her shoulders and sat knitting on the verandah. Their house had never been as lively since Will went overseas, but tonight it seemed like a tomb.

Johnny had gone directly to his room after his third piece of pie. Dada read the newspaper and Mima drank tea and watched him turn the pages. Before the war, Mima had considered it a shame that she'd never learned to read. When Dada would read the *Examiner* of an evening, she would watch him attentively, ask questions about the latest news. He would read whole sections of it to her aloud and they would discuss, argue and laugh about the foolishness of the world, as Mima called it. She never asked about the newspaper anymore.

Hattie pulled out her dressmaker's tape and measured the balaclava she was making for Will. Six more inches before it was done, and every inch seemed to take forever. Maybe it was the thin yarn and the small needles she was using, but she wanted it to be finely knit and warm.

Will had lost the first one she'd sent him and she'd started on a new one right away. But with the delay in the mails and how long it was taking her to knit this one, Will wouldn't get it until May, when it might be too warm for it.

"Hattie, are you going to bed?" Dada tapped gently on the door frame.

"In a bit. I want to get some more done."

"Can you see properly out there?" Dada was always concerned about her eyesight.

Hattie shrugged and turned up the flame on her lamp.

"Sure, I'm fine."

"Good night then."

When the light from her father's lamp disappeared upstairs, Hattie put her knitting in her lap and let her shoulders slump. It had been an agitated day. Johnny was the worst she'd ever seen him.

Johnny's hero worship for Will went over the top the minute Will came home in his uniform. But hero worship had quickly turned into hatred of the enemy and all things German. Following the lead of Thatch and Stook Hamilton, Johnny and Billy dug elaborate trenches, wrapped cloth strips around their calves like puttees and tied kitchen knives to wooden sticks for bayonet practice. They talked about dirty Huns, Jerrys, Heines, Turks, slackers, conchies and "doing your bit."

Johnny wasn't going to understand why their dad had David Ross on the farm. Heck, she wasn't sure she did. It was pretty clear, by the look in his eyes at the table tonight, that Johnny felt betrayed.

Darn that letter. She had told Johnny she'd burn it but she already knew she couldn't do that. Every letter and card she'd gotten from Will in the last two years was in that box. And every time she added another, it felt like a charm for his safety. If she burned that letter and something happened to Will, she'd never forgive herself.

She would move the box. But where? She couldn't put it in the house; her mother might find it. It had to be someplace special. Will's room in the barn was not possible, now that David Ross was staying there. There was only one place left.

In a clearing in the middle of the woods, there was a large flat rock. Left over from the last Ice Age, Will said. It had stood as Nottingham Castle in all their childhood games.

Every time the Sheriff was defeated and the prisoners rescued, Will, the undisputed Robin, would stand on the rock. His fists above his head, he would announce, "Freedom from tyranny, freedom for the peasants and justice for all."

And the Merry Men—Jimmy, Frankie, Herbie and Tom—would quaff pitchers of ale and roast large joints of venison.

Hattie was sure she could bury the box near the rock without anyone knowing about it. She tidied her knitting and stuffed it into the cloth bag she used to keep the yarn clean. Grabbing the lamp, she headed for the barn to get a spade.

Halfway across the yard, she stopped. A light suddenly appeared from inside the barn, flickering through the small window in the hayloft. But the light was moving. Gradually it dwindled from view, only to reappear, seeping through cracks in the barn door.

It must be David Ross going out to the toilet. Hattie braced herself. He would see her as soon as he opened the door. There was a haunting quality about him that made her nervous despite his quiet voice and gracious manners.

But the door didn't open. The light passed by the door then seemed to vanish into the interior of the barn. Hattie waited, but nothing happened. The night air crept into her clothing, chilling her arms.

"This is ridiculous," she said under her breath. It was her barn, her spade. She had every right to walk in there no matter what time of night it was. After what happened today, she couldn't leave the box where it was any longer. And she needed that spade.

Hattie put the lamp down outside the barn door. She knew Dada kept the garden tools just inside. Using both hands, she edged the door open a crack, then a little wider, a bit more, until the opening was big enough for her to slip through.

She waited a minute for her eyes to adjust to the light, then spied the tool she wanted a few steps to her left. To her right, on the far side of the barn and partially obscured by one of the beams, was David Ross.

He sat on a tall three-legged stool pulled up to Dada's workbench. His back to the door, his fingers flew along the workbench as if he were playing a piano. Hattie shuddered and turned to get the spade. She wanted to get out of there as fast as she could without drawing his attention. He didn't seem to pay her any mind, though, as she grabbed the tool and squeezed back through the door.

Retrieving her lamp and walking as quickly as she dared, Hattie crossed the yard and entered the dark safety of the woods. She carried the spade upright but had to untangle it from branches several times as she made her way to the cave.

Collecting the box wasn't difficult. Carrying the box, the spade and the lamp at the same time through the underbrush was more of a problem. At one point, she put them all down and tied her skirts up so she wasn't threatened with tripping at every step. Eventually she made it to Nottingham Castle Rock.

Using the tip of the spade, she outlined a section of sod, digging only until she hit the dirt below. Then she pried the sod loose, lifting it as carefully as pie crust and setting it aside. She dug a hole large enough to hold the box, carrying each shovelful of dirt into the trees and scattering it. With the box in the ground, she sprinkled a bit of dirt on top and replaced the sod. Carefully working the edges, she scattered leaves and twigs in places where the seams might be visible.

Only when she was finished did she hear the sounds that didn't sound like they belonged. The cracking of a branch that was too loud to have been made by any of the resident animals; a rustling in the leaves that was out of time with the wind.

Turning around with the lantern gripped in her right hand, she cast the meager light as far as she could toward the woods around the clearing but could see nothing out of the ordinary. So with her heart pounding faster than usual and her breath short, she took the spade and headed for home.

Hattie was thankful she'd tied up her skirts; it was much easier to tear up the path to the house. She stopped briefly each time she thought she heard crashing behind her. But every time, what she heard could have been the echo of her own steps. What made her scared was that she wasn't sure.

Breaking out of the woods, she pelted for the house,

recklessly swinging the lantern. Hattie crashed up the stairs and through the door, not caring what noise she made. The silence of the house filled her ears and she stood panting by the back window, still clutching the spade and the lamp.

As she watched the edge of the woods in the light of the quarter moon, Hattie saw a dark shadow, faint but tangible, emerge from the trees. It stopped, then moved quickly across the road and out of sight toward the river.

Hattie shivered and leaned the spade behind the door. He must have followed her. Her mouth went dry at the thought of what might have happened. He seemed so polite and gentle—could he have meant her any harm? She wished he'd never come, this David Ross.

September 1916

DAVID

*H*IS CHOIR ROBE FLAPPED wildly against his sides as David pumped his legs down the sidewalk. His father would be severe if David interrupted the new minister's first service with a tardy entry. Bracing one hand on the stone wall of the church, David pulled on the handle of the heavy oak door. His heart pounded and his quick breathing rasped into the stillness of the church. Thank God it was quiet—he was still in time for the first song.

Placing each foot flat on the steps to avoid squeaking, David climbed the twisting wooden stairs to the choir loft. His spot in the tenor section was open and he slipped quickly between the McClure brothers and smoothed the front of his midnight blue robe. He turned to give an apologetic nod to Mr. Liedermann, the choir director and organist. But instead, David's glance met the bulging eyes and disapproving scowl of Miss Agnes Willis.

The old maid sister of Mr. Willis, the church elder, had been the congregation's organist before Mr. Liedermann. She had ground out the same old hymns week after week until members of the congregation whispered decidedly un-Christian things.

So it wasn't long after Mr. Liedermann arrived, with his qualifications from the Vienna Conservatory, that the congregation drafted him into service as music director. The size of the choir had immediately tripled, and as the church filled each Sunday morning with extravagant, harmonious sound, Miss Willis had sulked in her brother's pew with a look on her face that Mother said would sour milk.

Miss Willis hadn't darkened the choir loft in years; what was she doing here now? And what was wrong with Mr. Liedermann? He hadn't ever missed so much as a practice, not even when his lumbago bothered him.

David didn't have time to find out. Miss Willis raised her large nose and gave the choir a knowing look. "One, two, three," she mouthed at them.

A chord rang out and the choir began. "Onward Christian soldiers ... " David scrambled through his book to find the right page. He noticed most of the older choir members didn't even crack the spines of their hymnals; they knew the words by heart.

David started to relax as the service continued. The new minister, Reverend Norris, had piercing eyes and a large well-kept mustache. His movements quick and purposeful, he strode across the sanctuary and opened the Bible as if he was marching a drill or loading a rifle.

"My friends, let us bow our heads this morning and pray for the men of the Canadian Expeditionary Force overseas. Soldiers both of Christ and of the Empire, crusaders like those of the First Crusade, bold defenders of the faith and of freedom."

David felt a shudder begin in the middle of his back between his shoulder blades. There was no doubt where this minister stood on the war effort, no doubt how he would feel about young men who didn't enlist. Drowning out the rest of the sermon with his own thoughts, David remembered the talk he'd had with Reverend Macauley last year.

The old man had sat, one hand resting lightly on his desk, in the manse parlor which doubled as his office. "What can I do for you, my boy?"

"I want to know, sir, what you think God meant by the sixth commandment."

"Thou shall not kill."

"Yes, but in what cases? Is it all right to kill some times and not others? And if it is, how can you tell which is which?"

"You mean like a soldier who kills the enemy in a war?"

"Yes, sir."

"How old are you, son?"

"Seventeen next month, sir."

"I see. Well. I've always taken the view that the commandments are absolute. Take the first one, for example. 'Love the Lord your God.' I don't think he could have meant love God sometimes, only on Sunday mornings, for example.

"On the other hand, I think God himself must make exceptions. The eighth commandment says 'Thou shalt not steal.' But if a man steals a loaf of bread to feed his starving child, I cannot think but that God would be merciful.

"The Bible says King David, for whom you are named, found favor in the sight of the Lord. He was also a warrior king and slew Goliath to protect his people from the Philistines. Are you worried, my son, that if you kill someone in battle, you will not be forgiven?"

"No. I... you see," David had whispered, "I don't think I could ever do it. Kill someone, I mean."

"Yes." The old man nodded slowly. "I'm not sure I could myself."

"But I want to do what's right."

"My son, the world doesn't always make it easy to know what God wants us to do. Pray. Read the Bible and you will get your answer."

"But what do you think?"

"My son, I cannot in conscience make a decision for you for which you will pay the rest of your life. The good book says, 'Seek and ye shall find, ask and it shall be given, knock and the door will be opened.' I can give no better advice than that."

David was shaken from his memories by the McClure brothers getting up, one on either side of him, to sing the recessional hymn. "God of our Fathers, holy faith. We will be true to thee 'til death."

As the last notes of the song drifted away and the congregation began to rustle in their seats, David caught the gaze of the organist. "Is Mr. Liedermann ill this morning?" he asked.

Her eyes, which always seemed to have a will of their own, flashed in the opposite direction as she leaned toward him in confidence. "Not ill, just put in his place."

Disturbed by her words and the smirk that went with them, David left abruptly and wound his way down the stairs, out the main doors into the brisk air. What did she mean? What had they done to Mr. Liedermann? Lost in his thoughts, he almost ran over Eloise Campbell on his way down the front steps.

"I'm sorry, Mrs. Campbell. I didn't see you. Are you all right?"

"I'm fine, David." Her glance at him was sharp, as if she wanted to ask him a question. Then he realized something—she wasn't in her choir robes and she hadn't sung with them today. That was an even rarer occurrence than Mr. Liedermann being absent. Mrs. Campbell had sung in the choir forever—since before Miss Willis was the organist.

As she turned to walk up Dovercourt Road, it occurred to David that Mrs. Campbell might know what was going on.

"Mrs. Campbell, may I walk with you?"

"You may."

David slowed his long strides to keep pace with her tiny, staccato steps. He wasn't sure how to start. "Choir was different today."

The small woman snorted softly. "Was it indeed?"

"I was late, you see, barely got there in time for the first hymn. And Mr. Liedermann wasn't there. And," he paused, "you weren't there."

She stopped abruptly, one small shoe connecting sharply with the wooden boards. "I'm as patriotic as the next woman, but Jesus said to love your neighbor as yourself—and Franz and Ada Liedermann are good Christian souls.

Driving them from our congregation is an un-Christian act, and you can tell anyone you like that I said it. Anyone at all."

It took a minute for David to take in what she'd said. "Driven out? You mean he's not coming back?"

One look at her face, clenched in anger, told him the answer. He felt a tightness in his chest. "But who would do such a thing?"

Mrs. Campbell's face softened. She laid one small gloved hand on his arm. "You liked him then? Mr. Liedermann?"

"Yes!" It surprised David how loudly the word came out, as if some force inside him had driven him to yell. "He's teaching me to play the piano."

"I see. This is sad for you as well, then." The fire and spunk seemed to have left her, David thought.

Removing her hand from his arm, she pulled gently on one gray-brown glove, as if to tighten the fit of the soft leather on her fingers. "But who did this? And why? Was it Reverend Norris?" David felt that he should know the answer to his question but it was camouflaged like an animal rustling in the underbrush.

Mrs. Campbell sighed. "He agrees, of course. It was why he was brought in by the session."

Her words seemed to fall like a marble down an empty shaft. The session—the committee of elders. It couldn't be—his father was on that committee. His father liked Mr. Liedermann, or liked his music at any rate. Mr. Liedermann and his wife had even been over for tea one Sunday afternoon before the war.

Before the war. That was it, then. This whole thing had to do with the war. David knew there had been grumbles in 1914 about having a German music director. Mostly, though, it had been Mr. Willis whom people suspected of motives other than patriotism. And Reverend Macauley had put a stop to it.

"We must not let so fine a thing as patriotism lead us to committing un-Christian acts," he had said from the pulpit. "Or we will become the barbarians we so abhor."

David looked at Mrs. Campbell. She must be nearly seventy but she stood straight as an arrow and every fold of her clothing looked as if it had been ordered to stand to. Wordlessly, he offered her his arm and they walked along the street together.

"Was it my father who did this to Mr. Liedermann?" he said at last.

"Your uncle is on the Methodist Army and Navy Board, isn't he?"

"Uncle Preston."

"McClure says your father brought him to the session. Your uncle said we hadn't been doing our bit for recruiting in this congregation. And with Reverend Macauley sick and too ill to continue, he offered to recommend someone who could stir up some national pride, encourage the young men to enlist.

"With talk like that, it didn't take long for someone to bring up the matter of our German choir director." She looked at him fiercely. "Though it doesn't matter who brought it up—it was unanimous except for Matt McClure."

They walked the rest of the way in silence. The petals on Mrs. Campbell's rose bush were scattered all over the grass from the rain the night before.

He bowed slightly at her front door. "See you next Sunday, Mrs. Campbell."

Her face became grave once more. "I don't think so, David. I'm an old woman and I've led a Christian life and the God I've been praying to all these years can hear me just as well from here as anywhere.

"Besides," she said, her eyes softening, "if they post recruiting sergeants at the doors, you never know, they just might nab me and whisk me off to France."

David smiled although his stomach was churning. Mrs. Campbell was kind, he thought, and didn't want him to leave on a sad note. He raised his hand in salute and left her surveying her rose bush, mumbling about how good she'd look in trousers and puttees.

DAVID WENT STRAIGHT to his music teacher's house, but a red-eyed Mrs. Liedermann met him at the door.

"Franz is out walking," she told him. "He left this music for you. Come for your lesson next week." She passed the sheet of music and a soft cloth bag through the door and retreated inside.

When he got home, David could tell they were having company for dinner. His little sister Katie was setting the table with Grandmother Tipton's Spode china.

"Is this your job today?" he asked.

Katie beamed. Usually Mother was the only one to touch the delicate plates.

Mother eyed him closely as he came in and for a dark minute he thought she might have found out about Mr. Liedermann's piano lessons. But it was just the egg he'd spilled on his jacket at breakfast.

"Tsk, tsk." She scraped at the spot; a pale yellow curl fell to the floor. She tapped his chest with the back of her right hand. "Take this off. May, see to this stain on Master David's jacket as soon as the table is set."

Setting down his valise, he struggled out of the jacket, sure his face was turning red. May bobbed in obedience, the way a maid was supposed to, but her eyes twinkled with mischief as he handed her the jacket.

His mother turned to check something on the stove and he slipped out of the kitchen and upstairs to his room, where he retrieved Mrs. Liedermann's gift from his valise.

Loosening the strings on the canvas bag, he pulled out what looked almost like a roll of bandages, tied together with thin strips of cloth sewn directly onto the roll.

Though his fingers fumbled, he loosened the ties and unrolled his present on the bed. At first, he didn't understand why Mrs. Liedermann would give him a quilted replica of a piano keyboard. Then he knew.

Pulling up a chair, he sat in front of the keyboard and tapped his right thumb on middle C as if he were playing. Hmmm. It worked. He heard the note in his head. He tried a G Major chord and heard that too. Examining it closely, he realized she must have made it herself. The white keys were soft flannel and quilted to give the sensation of a real key. The black keys were made of wool with more stuffing and felt just the right height compared to the white keys.

He dug out the Schubert impromptu Mr. Liedermann had left for him. His hands stopped trembling the minute he placed them on the keys. He played the first three bars, hearing the sounds in his head, even noticing when his fingers were misplaced. From then until May tapped on his door to deliver his jacket, David was lost in the music, working and reworking his fingers over the intricate patterns of the notes, his mind full of the imagined beauty of the sound.

The rap on the door startled him. His heart pounding fast, David realized he had to hide the keyboard before he answered.

"Master David?"

He rolled up the keyboard, pulled back the covers on his bed and hid the roll and the sheet music underneath.

"Master David, I have your ... "

He crossed the room quickly and opened the door. "Thank you," he said, taking the jacket from her arm.

She smirked a little. "Ma'am says to tell you that the guests have arrived."

"Who are they, May? The guests." He hated to ask, but May's lowered eyes and coy smile usually meant trouble for him, and he wanted to know what kind of teasing or embarrassment was coming his way.

"Why," she said, sounding surprised that he didn't know, "your uncle, Mr. Preston Ross, and ... "

"My cousin Julian," he finished her sentence.

"Quite so, sir." And with an exaggerated bob, May turned and tapped smartly down the stairs.

David sighed. Julian and Uncle Preston were the last people he wanted to see today. It was not, however, worth offending his father by avoiding the inevitable.

"There you are, David." His father stood in front of the stone mantle beside his older brother.

For a split second, David wondered who the young lieutenant was with his back to the door. Then Julian turned and David was forced to meet his cousin's steel blue eyes. Of course, he should have known! This was why they were here—to show off Julian's commission.

"Hello, David." Julian offered his hand with a smile. He was always pleasant and friendly in company; it was only when the two of them were alone that he was vicious.

"Julian has a commission with the PPCLI," David's father beamed. Oh well, he can be proud of his nephew if he can't be proud of his son, David thought. He knew his father was disappointed that David hadn't asked for permission to enlist before he turned eighteen. David was afraid to find out what his father would say if he told him he didn't want to enlist at all.

"And what is the PPCLI?" David's mother had Katie in tow.

"Aunt Marguerite!" Julian was always as sweet as four lumps of sugar to David's mom and little sister. "The Princess Patricia's Canadian Light Infantry."

"Oh my."

Julian offered his cap to David's mother. "You see the cap badge, Aunt Marguerite? It was designed by the Princess herself—a single daisy in honor of Major Gault's wife. And that's an odd thing. Major Gault's wife has a very lovely name." Julian paused, as if wanting them to ask for more.

Katie obliged him. "What is it, Julian? Is it Katie?"

"No."

"Is it Daisy?"

"No," he said, ruffling her hair, "it's Marguerite."

David's mother made a little choking sound and touched her lips with the fingers of one hand. "Oh, Julian, your mother would be so proud."

They all stood, quiet and uncomfortable, as David's mother composed herself. "Shall we go in to dinner, then? Preston?" she finally said.

David's uncle offered her his arm and they trooped across the hall to the dining room.

David summoned the Schubert from memory. He found, if he concentrated, that the music entirely drowned out the table conversation. The only exception was when his father asked him a direct question. But as his father's questions were usually rhetorical, a simple "Yes, sir" was a sufficient answer.

"Fine job you've joined up, Julian. You make us all proud, eh, David?"

"Yes, sir."

"Tailor did a bang-up job with the khaki. Looks very smart—Sam Browne belt and all, eh, David?"

"Yes, sir."

"Be your turn next, eh, David?"

David looked up sharply. It was Julian. To anyone else, David supposed, Julian's gaze would seem pleasant and interested. David knew differently. What he couldn't understand was how Julian always knew what he was thinking and how to target his weak points.

It seemed to David as if everyone was holding their breath waiting for his response. He would not give Julian pleasure by admitting his reluctance to enlist, but he also couldn't lie and build up false hopes for his father.

He decided to stick to a statement of fact. "I turn eighteen in July," he said.

"That's another five months." Julian's eyes sparkled. "The war could be over by then. You might miss all the fun."

David had spent hours over the years watching Katie's cat play with mice before he ate them. Julian must have learned some of his techniques from Old Tom.

"I don't think so," David said. "Bruce McLean said the war would be over by Christmas, enlisted right away so he wouldn't miss out on the fun. He's been over there three years already. So I imagine 'all the fun' won't be over by July."

"It has nothing to do with fun," Uncle Preston growled at David, his pale eyes glaring, his bushy eyebrows standing to attention. "We are all duty bound to do our bit. It is a question of honor."

David's father gathered his dinner napkin from his lap and set it beside his plate. "Almost a thousand young men signed up from the bank now, most of them with commissions. Almost feel sorry for the ones left behind. Young Jeffries, fine chap, his job's declared essential. He's just chafing to go over."

"That's the way it should be." Uncle Preston reached into his jacket pocket and pulled out his watch.

David's mother pressed her palms together. "It's time we went upstairs, Katie. Please excuse us, gentlemen." Julian was beside his aunt's chair in an instant, pulling it back as she rose. Katie peeked at Julian, her eyes full of what David called her "Lake of Shining Waters" look.

"Katie," David said quietly, "when is Miss Montgomery's next book coming out?"

She shot him a warning look. "You just be quiet, Davey."

David's mother paused in the front hall, smoothing her brown silk skirt with long slender fingers. "When do you sail, Julian?"

"On the sixteenth, Aunt Marguerite, from Halifax on the SS *Southland*."

"So soon? We won't get to see you again before you leave, will we?"

"No, ma'am, I expect not. It seems to take a lot of time to put a proper kit together."

Katie's eyes brightened. "Do you have a rifle?"

Julian chuckled. "Not yet, the army will give me that."

"Will you kill lots of Fritzes then?" Katie asked.

"Katherine!" Her mother and father spoke in unison and she froze to the spot.

Julian patted her arm and her shoulders slumped a bit. "Only as many as I need to, Katie." He glanced at David as if he was going to say something to him, then changed his mind.

David's mother offered her hand to Julian. "Our prayers are with you, son." She had started calling Julian "son" ever since the death of her sister, Evelyn, Julian's mother. David winced every time she did it.

Julian kissed her hand. "That's a charm against shrapnel. I'll be sure to take that to France with me."

David's mother ushered Katie upstairs; the brothers walked through into the parlor. Before he joined them, Julian lowered his voice and spoke to David. "I expect, you know, that killing Fritzes will not be much different than drowning unwanted kittens." Julian's smile showed all his perfect teeth.

David's roast chicken dinner began to roll and pitch, a carousel horse threatening to expel its rider.

THEY WERE BOYS THEN, eight and nine. Julian, a year older, was also bigger, much stronger and, it seemed, smarter. Well, no, not smarter, but he knew things, grown-up, dark and mysterious things that David could barely wrap his brains around.

David and his father were visiting Preston and Julian at their summer house near Orillia while his mother went to the Muskoka Lakes for her health.

Uncle Preston showed them a chaise he'd bought the previous spring.

"Lots of show wood here." Uncle Preston ran a gloved hand along one ornate curve.

"Be heavy for the horses to pull, I expect," David's father offered.

"They're beasts of burden, William. That's what they're for. Oh, cursed feline, look what's she's done!"

David climbed onto the step of the chaise and peeked

over the top. One of the barn cats, an enormous tabby, had delivered a litter of kittens in one corner of the seat. The kittens, blind and mewling, were spread out over the burgundy leather upholstery. They reminded David of Katie when she was first born. She was soft and delicate and smelled good, and he'd called her his little potato.

Uncle Preston swung open the door of the chaise, grasped the mother cat firmly in two hands and hurled her to the ground. "Get me a pail, boy," he growled.

Julian raced into the barn, returning with a shiny metal pail. Breathless, he handed it to his father. David felt confused but it was as if Julian knew what was going to happen next. His father's face was flushed and he seemed to be studying his boots intently. Am I the only dumb one? David thought.

Uncle Preston scooped the kittens up with both hands and dropped them into the bottom of the pail, which he then passed back to Julian. "You and Davey can take care of these, boy. You know what to do?"

Julian nodded, then beckoned to David. "Come on."

Unsure of what was to happen, but sure that he wouldn't like it, David glanced at his father. His father's face looked frozen, his eyes slightly puffy and pink, his jaw tightened, his teeth clenched. He didn't return David's look.

"'Way you go, lads." Uncle Preston rubbed his hands with a piece of gray flannel.

Reluctantly, David followed his cousin out of the barn, through the yard and the fence to the stream. Julian put the bucket beside the water and grinned.

"Do you want to do the first one?" he asked.

David wondered if Julian offered because he could tell David didn't know what was going on. He could always tell. "Are you ever stunned!" was his favorite line.

"No," David squeaked, "go ahead."

Julian reached into the bucket and pulled out a kitten, tight in his hand. He glanced at it, then plunged his hand, kitten and all, into the cold, fast-running water. He concentrated for a few seconds, then tossed the tiny body into the middle of the stream.

David's heart felt like a large hand was squeezing it. Julian dipped his hand into the bucket and retrieved another one, holding it out for David. "Your turn."

David's jaw fell open and he didn't have the strength to close it. Backing up quickly, he fell over a rock and landed hard on his bum. Not stopping to get up, he crawled backwards like a crab until he reached the fence, climbed through and ran as fast as he could toward the barn.

Julian's laughter nipped at his heels. "Sissy! Mama's boy!"

His father and Uncle Preston stood just outside the barn. But every step David took toward them, his feet seemed heavier, so heavy he could hardly lift them.

"Davey, what's the matter?" his father said.

David's foot smashed into a rock and he lost his balance, landing on his hands and knees, his face inches from Uncle Preston's boot. Looking up, he saw the big tabby peeking around the corner of the barn, her tail twitching.

Before he knew it was going to happen, he vomited. Not just once, but over and over, heaving until everything was gone. His father gave him a handkerchief to wipe his mouth and helped him to his feet.

"Overexcitement, I expect," his father said to Uncle Preston. His uncle said nothing, but David recognized the look on his face. It said, "Sissy, mama's boy."

THE SMELL OF SMOKE brought him out of his thoughts. His father had broken open his new box of cigars and lit one for Julian. That's a first, David thought. It must be the uniform.

"Spade doesn't sound German," his father said.

"Oh, he's German all right." Julian spoke with authority. "He's a carter over on Jersey Avenue."

David's father shook his head. "Seems senseless, to tar and feather someone like that. Lock them up if they're troublemakers, but ... "

"It was a gang of low-class ruffians that did it." Uncle Preston paused to draw a breath through his cigar. "But it's bound to happen, what with the casualty lists in the papers and these enemy aliens still living in the community."

David had to ask, though it meant risking Uncle Preston's displeasure. "They tarred and feathered someone just because they were German?"

Julian smiled slightly. "That's right."

July 1918

HATTIE

*H*ATTIE SLIPPED THE LETTER, unread, into her pocket and went back into the house. They'd heard two days ago that Will's best friend, Jimmy, was missing in action and right now, Hattie didn't want to face what might be in the letter. Her mother sat at the kitchen table hulling a mountain of strawberries.

"Get the water bath ready for them jars, Hattie." Her mother's voice was dull and flat, without its usual peaks and valleys of pitch and tone. Her hand poised above the strawberries, then picked one up and examined it. "Let's see, I pick one at random." Her voice quavered. She dug her small tool into the strawberry, deftly removing the stem. "I rip this out and throw it in the pan. I don't stop to think about who loved it or what it could have been if I hadn't picked it for my jam pot. Is that how it happens, do you think?" She picked up another strawberry and hulled it, throwing the stem onto the small green pile and the berry into the pot. "Is that how? With as little thought and attention as a woman hulling strawberries?"

Although it was just the two of them in the kitchen, Hattie wasn't sure her mother was talking to her. She was sure that even if her mother expected an answer, it was better to remain silent. Hattie stoked the fire in the stove, put the square tub on top and poured a pail of water into it.

Her mother sat, quiet now, holding a strawberry between thumb and forefinger, staring fixedly at it.

Hattie went down to the cellar, bringing up as many jam jars as she could carry. She made several more trips, lining the jars on the counter, but each time she returned, her mother was still staring blankly at the same strawberry. The water was starting to heat in the tub, but she'd used the last pail of water.

"Mima," she said, hoping to gently rouse her mother to the task at hand, "I'm going to get more water."

Her mother turned toward her but didn't speak. Hattie grabbed the two empty pails and swung out the screen door, down the path to the well.

Hattie loved pumping water, pulling the handle down hard, then up and down hard. Again. Up, down. Why did Jimmy have to go missing? Up, down. And why did Mima have to fall apart every time there was bad news? Up, down. And leave her to do all the work. Up, down. It was just like when they were kids and Jimmy got to be Little John and Hattie was left to mind the camp.

The pails were full too soon. Hattie felt like she could have pumped all day. Up, down, until the water poured all over the ground, soaking her shoes and making a river through the garden to the woods.

Instead, she lifted the heavy pails and trudged up the path to the house. The handles cut into her hands. Her shoulders stretched with the weight; it felt like her arms would pull off. Dada usually brought the water since Will went overseas.

Hattie took small steps up the hill, but each one seemed to bring a new thought she wanted to avoid. Step. What if Jimmy wasn't missing, but dead? Step. What if Mima didn't get better? Step. What if this war never ended, but went on and on until Johnny got his wish and became old enough to sign up? Step. It was too much; she set the pails down and rubbed her hands. Someone hurried up the path behind her. Will? She turned.

It was David Ross. "Can I carry those pails for you?" He didn't wait for her answer but scooped them up and strode easily to the house, Hattie hurrying to keep up with his long strides.

"Let me get the door." Hattie held open the screen door and followed him in.

"Your father sent me to ask you for a bag lunch. We'll be moving to the far pasture, and he says it's too far to come back at dinnertime.

"Oh!" With all the fuss about the jam, Hattie'd forgotten about dinner. "Mima?" She was gone. The strawberries were still piled on the table, only a few in the pot.

"Mima?" Hattie almost screamed, an edge of hysteria creeping into her voice. She checked the parlor, found it empty and returned to the kitchen.

David Ross looked at her. Damn those eyes of his that seemed to know everything. "Would she have gone to lie down?" he asked quietly.

No, was her first thought, not with all this jam to make. Mima would never do that. But Mima wasn't acting normal anymore. Her worry over Will seemed to get bigger and bigger as the war went on, until her head couldn't hold anything else. Hattie left David without a word and ran upstairs to her parents' bedroom. It was empty. She felt something large slide down her throat, lodging in the pit of her stomach. Where would Mima have gone?

Hattie almost missed it; the boys' bedroom door was only ajar. But it caught her eye because she'd made sure to close all the upstairs doors before she came down to breakfast. She pushed it open.

Her mother lay across Will's bed, one strawberry-stained hand resting on the crocheted edge of the white pillowslip. Hattie stayed long enough to make sure her mother was breathing. She never stirred when Hattie draped the throw from Johnny's bed over her.

How would she ever make all that jam by herself? And supper too? Right, and lunch for the men. David Ross was still waiting for it downstairs. She closed the door behind her and trotted downstairs to the kitchen, trying to remember what was in the pantry for a fit lunch.

Her mind on pork pie and raisin cookies, she didn't notice at first that David Ross was hulling strawberries.

Hattie, her arms full of pork pies, stared as he hulled one after another, lickety-split. He glanced up and smiled slowly, never breaking the rhythm of his hands with the berries.

"My grandmother made jam in the summer and this was always my job. I would race with myself to see how many I could do in fifteen minutes. The clock in the hall would chime every quarter hour and I would start over, trying to beat my best total."

His hands were like graceful birds, darting over the strawberries, separating them from their stems with the slightest of movements. Hattie thought she noticed the pile of berries on the table already getting smaller.

By the time she assembled and packed a big enough lunch for three and made a jug of tea, the strawberries were half done.

She almost thanked him for his help but caught herself in time. He was here, safe, hulling strawberries and splitting rails while Will was in some lousy trench—facing mortars and bombs and snipers that at any moment might kill him.

After David left, Hattie put the jars in the water bath and sat down to finish hulling the strawberries. But what her mother had said earlier came back to haunt her task. She could hardly hull a berry without feeling somehow that she was bringing bad luck to Will and to Jimmy, wherever he might be. Unless, of course, she thought, the strawberries were Germans.

This made the job easier. She imagined each berry as one less soldier on the Hun's front line. Jerry berries. Hattie sputtered a nervous laugh. She could even put it on the labels—German Strawberry Jam, 1918. She'd show that conchie, David Ross.

It was late evening before Hattie got all the jars filled with jam and sealed with wax. Her mother got up only once, had a cup of tea and went back to lie down, this time in her own room. Hattie made supper for everyone and did the washing up before getting back to the jam.

So it was dark out and the full moon rising when she lined up all the jars to count and label them. She'd always thought this was the best part of home preserves—the pride she felt at the sight of the freshly filled jars, warm and labeled, ready to be put in the cellar. Tonight, though, her pride was mixed with shame. The jokes she'd invented while she made the jam made her sick. Jerry berries! How could she wish death on people she didn't even know? She'd never felt that way before. Maybe she was going crazy with fear like her mother.

Hattie hung her apron on a hook, grabbed her sweater and went outside for a fresh breath. If only she could do something. Work in a factory in the city, even sign up as a volunteer nurse like Jimmy's sister, Agnes. Then she'd feel like she was really doing something—not just stuck at home, doing more and more of her mother's work while Mima faded into a shadow.

If the moon hadn't been full, she would never have seen David Ross head for the woods. So it had been him she'd seen that night. What could he be doing? Captain Nelson had said that conchies were often spies. Up to now, Hattie had thought it just one of his crazy ideas. But what other reason could David have for going to the woods this time of night? Hattie determined to find out.

She was glad for her Sherwood Forest training. It all came back to her, slipping through the trees in her stocking feet, her skirts tied to one side, her laces knotted and the boots slung over one shoulder. She knew every sound of these woods and could hear every step David Ross made. He stopped somewhere near the center of the stand of trees—somewhere near Nottingham Castle Rock. It was a natural place to meet someone—if that was what David Ross was doing.

She saw his shadow near the rock and crept as close as she dared, beside the bush from which they'd always launched their attacks on the castle. It's a good thing this shrub has grown, she thought, because so have I.

David Ross pulled a piece of log beside the rock and sat down on it. He reached inside his jacket and removed something. Then he unrolled it onto the rock—it was a cloth piano! She remembered the night she had seen him at the workbench in the barn. He sat for a moment, his hands poised above the keys, his eyes closed. Then he began to play. His head nodded and his body swayed as his long thin fingers danced up and down the cloth as if it were an ivory keyboard. Hattie noticed that he also moved his right foot every once in a while as if he were pressing on pedals.

Hattie knelt in the grass and held her head to keep it from spinning. She was all alone in a world gone stark raving mad. Will and all his friends were gone to France these two years. Johnny followed Thatch and Stook Hamilton all over the county, mad to kill the Hun, and her mother was paralyzed with fear. Her father worked from dawn to dusk, and this one, this interloper, this cowardly stranger, stayed up half the night pretending to play the piano in the woods. The rest of it she couldn't change, but this one, this one she could do something about.

Creeping around behind him, Hattie slipped her boots from her shoulder to the ground and sprang into a run. She snatched his pathetic cloth piano out from under his fingers as she ran around to the other side of the rock.

At first his look was startled, then puzzled, as he gaped at her. "Hattie?" he finally said.

She glared at him while she rolled up his piano and tied it together with the ties she assumed were there for that purpose.

His eyes softened and he smiled faintly. "Was it too loud?"

Hattie breathed very hard through her nostrils and tightened her lips. She would ignore his stupid joke.

David Ross stood up and held out his hand.

"Hattie, may I have it back, please." He spoke softly, but Hattie thought she caught a hint of desperation in his voice.

"No."

"Why not?" His voice was a whisper, a small rustle of wind. "This hurts no one."

"Hurts no one?" Hattie's voice climbed. "Is that your motto? And what do you know about what hurts? My mother," she gasped, "my brothers, one in danger every moment, the other twisted by blood lust. And you, out here, crazy as a loon, playing with a piece of cloth. You think it doesn't hurt? Everything hurts. You should be in an asylum."

Hattie turned away and hugged her arms tight to her chest. Her throat ached and burned, but she wouldn't let this coward see her cry. She took a deep breath and held it for as long as she could. And another and another.

It seemed to her a very long time of complete silence before she felt a light touch at her elbow.

"I'm practicing. It's an impromptu by Schubert. I have the music in my room. I can show you if you like."

Hattie pulled her arm away, her mind fighting against the reasonableness of his explanation. She didn't want him to be right; she wanted him to be crazy so he'd have to go away. Maybe it wouldn't make any difference, but maybe Johnny wouldn't be so wild if there wasn't a conchie about the place. "Owning a piece of sheet music doesn't mean you can play it."

"I'm sorry, Hattie. The only way I can prove it to you is with a piano."

She looked at him sharply. He knew they didn't have a piano, and smart little slacker that he was, he probably thought there was no chance of her coming up with one in the middle of the night.

"All right then," she said, striding around him to retrieve her boots. She ignored him as she struggled to get them back on and laced them up to the top.

"Let's go," she said finally and headed west, away from the farm.

"Hattie, where are we going?"

"To Mrs. Nelson's house. She's got a piano."

"You're going to wake her up at this hour just to prove I'm crazy? Hattie, you worked too hard today."

Hattie slapped at branches that weren't even in her way. "We won't wake her up. She's in Kingston with her sister. What's the matter? You afraid?"

WHEN THEY GOT to the Nelson place, a tiny bungalow in a clearing by the river, Hattie made David Ross wait at the door. Mrs. Nelson had warned her never to show anyone else where the key was hidden.

Hattie followed the strand of red yarn to the end of the chicken wire fence and pulled on it. A small cloth bag tied with string popped out of the dirt. Hattie undid the tiny bow and shook out a brass key.

Dada had always called Mrs. Nelson a "complicated" woman, and Hattie was beginning to understand why. Anyone else in the county who bothered to lock their doors would have put the key under the front mat or a flower pot. Mrs. Nelson was both elaborate and secretive.

Returning to the porch, she unlocked the front door. The smell of old lavender rushed out at her. With all the shutters closed, it was as dark as an Egyptian tomb. So Hattie felt her way through the parlor to the dining room table. The lamp was there and the neat silver matchbox Mrs. Nelson always set beside it.

Striking a match, she lit the lamp and turned up the flame. What she saw amazed her. The table, the sideboard, the buffet and every small table in the parlor were covered with vases. Mrs. Nelson must have picked every flower in her garden and arranged them like this before she left.

Long dead and dry from the summer heat, fallen petals lay scattered everywhere. Mrs. Nelson must have known this would happen. It was so unlike her to leave a mess that it made Hattie quite nervous. Perhaps it meant she wasn't ever coming back.

A note sounded on the piano. Startled, Hattie turned. David was completely absorbed with his examination of Mrs. Nelson's upright piano and hadn't noticed her discomfort.

Sitting on the bench, he caressed the keys gently without making a sound. Then, straightening his back, he began to play scales with both hands, up and down the keys, faster and faster.

The picture of Captain Nelson on top of the carved oak piano teetered as if it was going to fall. Hattie started to speak but David Ross had noticed it too. He deftly plucked the frame to safety before Hattie could utter a sound. Turning, he placed it gently on the nearest table, in front of a vase of dead lilacs. His eyes met Hattie's with enthusiasm and warmth, and she was taken aback.

He attacked the keys again, this time playing music Hattie had never heard the like of. Graceful and light, his fingers flew all over the keyboard. Trilling and patterning, the music seemed to fill the little house, shaking all the cobwebs from the walls.

Sinking into Mrs. Nelson's softest parlor chair, Hattie admitted how tired she really was. She closed her eyes and let the sound wash over her and carry her away like a leaf on the river, downstream, cradled in the soft hand of the water. As she sank back, her thoughts whirled, as if pulled down in an eddy into the place she'd been avoiding all day.

Hattie had been jealous of Jimmy for as long as she could remember. Every time she'd felt left out when Will's friends were around, it was Jimmy she blamed. Eventually she'd realized that her fury was aimed at him not because he was the worst of the lot, but because he was Will's best friend. Grudgingly, and only to herself, she admitted that aside from Will, Jimmy was the brightest and the best of the Sherwood Band. But she remained jealous.

Will and his friends enlisted together as soon as school let out in 1916. But two days before they were to leave, Jimmy came down with the measles. The rest of them left on the train as planned; Jimmy had to wait, first to get over the measles and then for the next muster to be organized.

He'd asked Hattie to see him off at the train. In fact, he'd made excuses to come over almost every day after his spots cleared.

"You're not going to hate me for the rest of your life, are you, Hattie?" He'd smiled a wicked, teasing smile. "I'm not contagious anymore. Come for a walk with me."

"Mima needs help with the washing up."

Hattie's mother had snorted. "Now don't be using me as a poor excuse. You go with Jimmy."

They'd walked through the woods to Nottingham Castle Rock.

"Are you still handy with a quarterstaff?" he'd asked.

"Better than you."

He set his mouth in a grim line. "I hope you're wrong. I have to be at the top of my form over there."

Hattie hadn't known what to say. Their childhood dreams of adventure had become nightmares, and once again, she was being left behind.

So when Jimmy had asked, "Will you come to the station with me? See me off properly?" she had agreed.

The scratch of green wool on her cheek, the hiss and the smell of the locomotive, merged, then sank beneath the chords of the music.

She woke with a start, not sure at first where she was. David Ross stood quietly in front of Mrs. Nelson's piano.

"It's past midnight," he said.

Hattie jumped to her feet. Her mind was groggy but she quickly determined that it would not be wise for them to go back together.

"You go first. I have to … " she looked around the room for an excuse, "clean up these flowers."

David Ross nodded and turned toward the door. Smoothing her skirts, Hattie felt the bulge in her apron and remembered what had brought them here. Glad he could not see her flush in the dim light, she removed his cloth piano.

"Mr. Ross," she said, holding it out to him. "Your music is beautiful."

"It was a gift to play on a piano again." He smiled and retrieved his property from her outstretched hands. "Thank you."

Hattie wanted to apologize but she couldn't. Not without talking about Mima and she couldn't do that. So she gathered up vases of dead flowers in her arms and let David Ross stride home in the moonlight without an explanation.

She emptied the vases, piling the flowers at one end of Mrs. Nelson's overgrown garden. Hattie had meant to come before this and weed the vegetables but with all the extra work at home, she hadn't managed the trip. She swept up the petals from the floor and the furniture and was about to go when she saw the photograph of Captain Nelson still on the table.

She restored it to its place of honor on the piano, although the last time she'd seen Mrs. Nelson handle the picture of her dead husband, it had been with anger, not love or respect.

"Look at this photograph," she had said to Hattie one spring day as they sat together at the kitchen table. "His tidy mustache, his uniform tailored to order, all his leather and brass polished to a shine and his puttees wrapped to regulation. It would make you think he had a brain, wouldn't it?"

Hattie, astonished at Mrs. Nelson's angry tone, nodded gravely. Mrs. Nelson had always seemed so proud of her husband.

"Well, he didn't and that's the truth. His cousin, Edna, ever the kind soul, when she heard the bad news, packaged up all the letters he'd written to her from overseas and sent them to me. She thought, I suppose, that they would be a comfort to me. Hah!

"Listen to this—'*Dear Edna, Nice to hear from home, etc. etc. etc.*' Here it is.

> '*Nights in the trenches are so boring. The only occupations seem to be pinching lice, shooting rats and trying to get an hour of sleep. To be honest, sometimes to cure the boredom, I play a little game with Jerry. I poke my head above the parapet, just an inch or two. Then, quick as a wink, I duck back down. Just a bit of fun to pass the time.*' "

Hattie had been stunned. Everyone knew that Captain Nelson had been killed by a sniper's bullet to the head.

Mrs. Nelson had grabbed the pile of letters and strode purposefully to the stove. Lifting one of the lids, she stuffed them one by one into the firebox. "All those months of worry, knitting socks until my fingers bled, waking in the night drenched in sweat from some nightmare about him, all for nothing. Because coming back in one piece didn't matter any more to him than a foolish game. I am ashamed that I bothered to worry at all." She took the last letter, ripped it into tiny pieces and let them fall slowly into the flames.

Hattie had quickly finished her tea and made an excuse to go home. But she wasn't surprised two weeks later when she heard Mrs. Nelson was spending the summer with her sister in Kingston.

Now, she remembered the icy look of determination on Mrs. Nelson's face as she'd burned the letters. Hattie couldn't imagine burning any of Will's letters, not even if the worst happened.

Will's letter! It was still in her pocket—she felt ashamed that with all the commotion she hadn't thought of it all day. She wandered back into the dining room, set the letter beside the coal oil lamp and eased herself into one of Mrs. Nelson's hard-backed oak chairs. Rubbing her finger along the envelope, she tried to imagine what was in it. Part of her didn't want to know.

His hand was unusually sloppy and large, as if he had written under stress or in a hurry.

My Dearest Hattie,

I don't know even how to write this to you. Jimmy's hinted a lot over here. Things like, "Hattie doesn't hate me near as much as she used to" and he'd introduce me as "Will Tamblyn, the man whose sister saw me off at the train."

So I don't know what to say except that if the two people I love most in the world have found some happiness with each other, then I couldn't be more pleased.

And this is why I hate myself, because Jimmy's missing in action, and I'm left to tell you. Three nights ago, we were getting ready for a show. Jimmy and two other fellas went out on a forward patrol. Then the show started and nobody's seen anything of them since.

As I write this I keep half expecting him to lope into the billet with that tease of a smile on his face. So don't give up hope, as he may have been wounded and sent down through a different dressing station or even captured. Please God, I will have better news when I write to you again.

Love, your brother Will

Closing her eyes, Hattie caught the scent of wool—Jimmy's greatcoat on her cheek as he pressed her to his chest before he boarded the train. It had all been so sudden and strange, she hadn't been sure what she felt. She still wasn't.

Hattie folded the letter and slipped it back into her pocket, lifted the glass and blew out the lamp. She would lock up the house and put the letter in the box in the woods before she went home.

Almost as soon as she entered the woods, Hattie thought she heard footsteps behind her in the dark. Once she even called out, "David Ross?" but there was no answer.

At the rock, she felt in the cool grass for her hiding place. One hand reached under the sod and lifted the lid of the box, the other slipped the letter inside. Grabbing her skirts in both hands, she ran for home along the familiar path, trying to remember where roots and branches had to be avoided.

The closer she got to the yard, the louder the crashing behind her seemed to get. Her mouth felt dry and sticky. How could she have been so stupid as to go into the woods alone with him? She'd been deceived by all his false charm. His quiet manners and the gentle looks he gave her lulled her into security. Then he waited in the woods to scare her as she came home. That made her angry. What made her scared was wondering what he'd do if he actually caught her.

The shadows of the trees ended just ahead. If I can only get to the house, she thought. Damn. Her right boot slammed hard against something that wouldn't give way. The rest of her plummeted past and she fell nose first in the grass at the edge of the yard. The heavy crunch of boots and someone breathing heavily through the mouth came up behind her.

From near the barn, a lantern swung toward her. "Dada?" Her voice squeaked as if it had lost all its strength. The creature in the woods faded away as the light came nearer and nearer.

The lantern was set in the grass and strong hands grasped her from behind under her arms and lifted her up.

"Can you stand?"

Her head snapped. It was not her father, it was David Ross. Her sudden movement made her teeter on her painful right foot.

David encircled her shoulder with his arm and gripped her firmly. "Just lean on me for a minute. Make sure you're okay before you try to move."

Hattie took short breaths and swallowed, trying to find her voice. If David Ross was here, in the yard, then who had chased her through the woods? Maybe she was going crazy.

"My foot hurts," she told him. "I walloped it on a rock or something." Tentatively, she put her weight on both feet. The right one ached, but she was sure she could walk on it. "I think it's okay."

Releasing her shoulder, he took her hand instead. "Try a few steps."

Gripping his hand tightly, she limped toward the lantern a few feet away. "Did you... hear anything? In the woods, I mean."

"I heard you coming. Were you running?"

"I don't mean me. Did you hear anything else?" She stood still but could not bring herself to let go of his hand.

"No. What was it?"

"I thought... there was someone else in the woods." She searched his face as she spoke. "I thought it was you." His eyes looked soft in the lamplight and he shrugged slightly. "It's happened before," she added.

His brow furrowed. "Could it have been an animal?"

"It was too big for that and it didn't feel like an animal. It felt... human."

David Ross stroked her hand then released it. Picking up the lantern, he gestured toward the house. "I'll wait here until you're safely inside."

Watching from the window, she saw David Ross head for the barn, drifts of light spilling into the yard as he climbed up to his room.

David lay on his back, eyes closed, fingers laced behind his head, the impromptu still racing through his brain. The way she'd pressed his hand had every nerve in his body humming. It had been a very long time since he'd played on a real piano.

May 1911
DAVID

THE SATURDAY BEFORE THE Ladies Aid bazaar, David's mother asked him to help with the setup in the church basement. When the tables were in place and the ladies descended on them with home preserves, pies and doilies, David escaped upstairs.

Music like he'd never heard was being played on the old pipe organ. David sank back in a wooden pew and closed his eyes. Was this what they meant by the glory of God?

There'd been a piano in the parlor at home as long as David could remember. An upright Heintzman, dark oak with a matching stool. The cover was always closed, draped with a piece of lacy crochet work. He remembered precisely the first time he heard it played, though his mother said he must have been only two years old.

Waking from an afternoon sleep, he'd heard unusual sounds from the grown-ups' room. His mother sat on the stool in front of the funny table, her fingers dancing across blacks and whites—sounds that made his ears happy. He'd climbed up beside Mama, and she'd shown him how to touch the keys to make the sounds.

After that, every afternoon, David threw the lace doily on the floor, pushed open the lid, crawled up on the stool and made sounds on the blacks and whites. Sometimes Mama would sit beside him and play too.

It must have been on one of those occasions that Father came home unexpectedly and put a stop to David's early musical education. David didn't remember a particular incident, but he did remember a day when the parlor doors were locked. From then on, afternoons were spent taking walks instead. He grew to understand that music was not something boys did. Boys did athletics and studied mathematics so they could become businessmen and bankers.

He joined the track and field team, specializing in the longer distances—the ones that required endurance rather than speed. And although he produced firsts in math all through school, he did not have his father's love of numbers. Sometimes, watching his father's face, intense and bright as he added large columns of numbers without ever making a mark on the paper, David longed to share his father's passion.

By the time David was twelve and old enough to help set up the tables for the Ladies Aid bazaar, he had stopped remembering the afternoons with the blacks and whites and had resigned himself to the life of a reluctant banker. The music on the church's old pipe organ, mysterious and stirring, made him remember what it felt like to play music. When it stopped suddenly, he climbed the loft stairs, not sure what he would find. He heard muffled grumbling but he couldn't see anyone. Moving slowly toward the organ, David saw a man, his mostly bald head fringed with closely cropped white hair, on his hands and knees in front of the organ. One of the floorboards squeaked and the man turned, catching sight of David.

"Ah, God has sent me a helper. Come, come."

Although it seemed a little strange, David was too curious not to do as the man asked. On the floor beside him, the man held two pieces of the pedal together and fingered a notch at the top of the pedal.

"For this job I need three hands, perhaps four. These pieces must be held tight together, so I can fit this peg into the notch. Like this. Can you do it?"

David squeezed the boards together.

"Wonderful."

The man stood up and fired out his right hand. "Thank you..."

"David."

"Thank you, David. I'm Franz Liedermann." He indicated the organ. "Do you play?"

David shook his head, but he reached out and pressed a couple of keys. "I've never heard music like what you were playing. What was it?"

The man's bushy eyebrows raised a little. "Bach—the *Contrapunctus*. Not written for church. But Bach was close to God, yes, so even his studies in counterpoint sound somewhat like a prayer."

David wanted him to play again but thought it might be rude to ask. So instead he asked, "Are you the new organist?"

"Ah!" Mr. Liedermann smiled. "Perhaps it will happen. Mr. McClure has put in the good word for me." Digging in a leather valise, he pulled out some sheet music and arranged it on the organ.

"You can sing, yes? How about helping me practice?"

David wasn't sure the singing he'd done in primary school counted, but he wanted more than anything to hear Mr. Liedermann play the organ, so he nodded.

Mr. Liedermann struck the paper sharply with his index finger. "Bach, as well. The *Cantata 147*. We will sing the chorale; this is a famous piece, yes. And then, perhaps, you can try the second movement—the recitative which Bach intended, well, for a boy about your age."

David was sure confusion showed on his face because Mr. Liedermann waved his hand.

"No matter—we start simple. I will play it first so your brain can learn the sound."

The first thing David noticed was that the singing leaned against the music played on the organ instead of dancing on top of it. Mr. Liedermann smiled with all his teeth when David tried to explain it to him.

"Cooperation, yes! That is what Bach is all about—two separate lines that balance vertically." He waved his hand at David. "You must have an affinity for Bach. Men study for years to understand what your ears have picked up the first time."

They sang for two hours, Mr. Liedermann showing David how to stand and to relax the throat muscles to allow for proper airflow. He tapped David's Adam's apple.

"This is a musical instrument, you must hold it properly. Feet shoulder-width apart, yes. Now put your arms in front of you, hands together, thumbs up. Open your hands, thumbs out. Good. Arms to your side, now down. This leaves your chest out and your back straight so the air can flow through the instrument." He tapped his own Adam's apple. "See what happens."

Straddling the piano stool, Mr. Liedermann hunched over so his beard rested on his chest and began the chorale.

"*Wohl mir, dass ich Jesum habe,*" he croaked.

David laughed out loud. The old man smiled and nodded. Fishing in his pocket for his watch, Mr. Liedermann began to pack up his music.

"Ada will wonder what happened to me." He offered David the sheet with the chorale on it. "Would you like to learn to read the notes?"

It was out of his mouth before he had time to consider his words. "I want to learn to play but Father would never allow it."

"Hmm." Mr. Liedermann ran his fingers through the underside of his beard. "We will figure it out, yes."

January 1918
DAVID

MR. CAZEN, THE ROSS FAMILY solicitor, clamped his narrow lips into a grim line when he heard the decision of the tribunal. He tidied his papers and ran his hands over his head, flattening the wisps of hair that had come loose during the hearing.

He touched David's arm lightly at the elbow. "Don't worry, we have one more appeal." The lawyer's face fell momentarily into a grimace. "But I wish you'd consider the employment exemptions. I'm sure that even at this late date, I could arrange for you to work on a farm. Much easier to get through."

David shook his head slowly. "I'm not a farmer, Mr. Cazen."

"You don't have to be one, son. Just be willing to work on one." The solicitor's voice was gentle, his eyes a mix of curiosity and compassion. As the guards came to take David away, he spoke again. "You think about what I said now. It'd save you a lot of trouble. I'll see you in the morning."

He was grateful for Mr. Cazen's discretion. Although David knew that he worked under the direction of his father and Uncle Preston, the lawyer never once alluded to it. Never once did he act as if David wasn't his sole client.

Even about the farm exemption. He'd said, "It'd save you a lot of trouble," not "It'd save your family a great deal of embarrassment." David imagined his uncle yelling at Mr. Cazen and knew that the lawyer's kindness to David was not without cost to himself.

Back at the depot, the sergeant seemed unsure what to do with him. He left David sitting on a wooden chair in the hall while he discussed the problem with someone in an office. David wasn't sure if the door was left open on purpose or not.

"So, what do I do with this high-class slacker then, sir?"

"What do you mean, Sergeant?"

"Well, I can't just throw him in the hole with the rest of them, can I?"

"Why not?"

"Well, sir, slacker or not, he is a Ross."

"And?" The tone was sarcastic.

"His uncle single-handedly outfitted an entire regiment and headed up recruitment for the Methodists. What would he think if we treated his nephew like any other skivvy?"

The officer sighed. "What do you suppose he thinks of his nephew, the conchie? Do you think he cares about this prize shirker while his only son is on the line in France?"

The sergeant sounded gleeful. "In the hole then, sir?"

"Maybe it'll open his eyes to the kind of low-life and riff-raff he's associating with."

The big man grimaced and pointed down the hall as he emerged from the office. "Down there." David rose at once and followed the sergeant. At the end of the hall they went down winding concrete stairs into a dank-smelling basement. Large windowless metal doors lined the hall on either side.

The sergeant grabbed David's arm as they reached a door halfway down on the right. He dug out a key, opened the door and pushed David into total blackness. The rush of air as the door closed behind him forced the smell of the room into his nose. Urine and smoke.

David froze. Not sure what was beneath his feet, he stayed where he was and strained his eyes to see in the blackness, without success. Mumblings from the far end of the room told him he wasn't alone, but he couldn't make out the words or the speakers until one of them lit a match.

"Welcome to the hotel of conscience, dear traveler," the figure said, using the match to light a candle. "Apologies for leaving you in the dark at first, but unfortunately these candles, like our objections to the war, are considered illegal by Sergeant Wilkes and company." He waved his arm. "Come on over and introduce yourself."

DAVID WALKED SLOWLY toward the group, trying not to think about the wet floor he trod on. The face reflected by the dim light was small and grimy. The smile revealed jagged blackened teeth, and the eyes were bright and shone with humor.

"So, what kind of a CO are you?" the man with the candle asked.

"I don't know what you mean."

The man motioned for him to sit down, but when he

hesitated, the man said, "It's dry over here." He smiled his broken-toothed smile as David eased himself down onto the concrete.

"I'm Pete, the socialist kind. The first time the medical officer looked at me, he said I was D, for death warmed over. When they found out how many workers I'd signed up for the union, old doc seemed to change his mind. Said being toothless, blind in one eye and consumptive made me prime material for the front line trenches. So here I am, objecting."

"Johann and Ernst over here," Pete said, waving to the two young men next to David, "are the Mennonite kind."

"I thought it was automatic exemption for Mennonites." David was glad Mr. Cazen had gone through the rules with him carefully.

"It's supposed to be," the blond one called Ernst said, "except we must prove that we are Mennonites. We are only twenty and in our church, we do not become baptized until we are twenty-one. The officer would not accept the word of our elder because we are not on the church roll."

Pete chuckled. "So they're objecting. John Henry over here is a Christadelphian. He's almost forgiven me for thinking that means he is a Christian from Philadelphia.

"Thomas and Simon are friends—real ones—from the Society of Friends."

Simon nodded at David. "Quakers. Not everyone knows the official name."

"Henri," Pete continued, "is from Three Rivers and objects to being dragged into an English war. Bob and

George are International Bible Students, and that's the whole objectionable lot of us. Except for you." David let the last words hang in the thick air. The faces around him were all quietly expectant, but he was unsure of what to say. In some ways it was so simple. He just couldn't kill anyone. In other ways it was so complicated. He couldn't hate the Germans. Mr. Liedermann had given him the gift of music, something his father had said was not worthy of a man's pursuit. He'd struggled over and over to understand why it was so easy for some people to hate with such vehemence and certainty.

He remembered Mrs. Smith approaching his mother after church, saying it was unpatriotic to listen to the music of the German composers. His mother, with consummate grace, had quietly dismissed her and said privately to David that because Mrs. Smith had no husband or son to offer up to the cause, her only notoriety must come from finding new and inventive ways to hate the Hun. David wasn't sure, however, that Mrs. Smith was alone.

Henri dug in his clothing and pulled out a crumpled cigarette. He leaned over the candle, sucked the flame toward him until the end of the cigarette glowed red. Mrs. Smith hated the French Canadians too. She claimed to have written the Prime Minister, telling him that conscription should only apply to French Canadians because the men of what she called "the Lower province" had not done their share. She would pull out Mr. Borden's reply for anyone who showed the faintest interest.

David had never understood Mrs. Smith. He looked now at the faces around him. He wondered if they had the answer, if they had searched in places he had never looked and found what would lift the heavy weight pressing on his chest.

"My name is David. I just told them I couldn't kill anyone. That it was against my conscience." David almost whispered the words.

"Good lad," Pete said.

Simon reached across the circle and shook his hand. In this dank musty room, Simon's flesh was warm and dry. "What faith are you?"

"Methodist."

Henri snorted. Pete laid a hand on Henri as if to restrain him. "Now then," he murmured.

There was a moment of silence. Henri took a long drag on his cigarette. His dark eyes met David's. "I'm sorry. You see ... " He gestured with his hand, small flakes of ash falling as he spoke. "... where I come from, Methodists are not exactly pacifists."

"No," David said, "not where I come from either."

Henri grinned, tapped his index finger on what was left of his cigarette. "And here we are."

The sound of a key rattled in the door. Pete spit on his fingers and extinguished the candle in one quick movement. They waited in silence as the door opened. Two pails were set down just inside and the door slammed shut.

Pete waited several minutes before relighting the candle. Ernst and Johann got the pails. Ten hard rolls, one for each of them, and a pail of water. Pete broke his in half and soaked it with water. David watched him scoop mushy pieces into his mouth.

"Can't chew much of anything," he explained. "It'll crack the few teeth I've got left."

David cupped his own roll in his hand. He wasn't hungry. Mr. Cazen had insisted on eating a big dinner before the hearing—crown roast of pork with applesauce, mashed potatoes, gravy and bread pudding with lots of raisins for dessert. He felt conspicuous as the other men devoured the dry, unappetizing bread. Johann finished his first, then laced his long fingers together in his lap. He was tall, six foot four maybe and muscular from hard work, David supposed.

David held out his roll to Johann.

"No," he murmured.

David pushed the bread toward him.

"Take it," Ernst said. "You know you're hungry." Ernst picked the roll from David's hand and dropped it into Johann's lap. "We're cousins. You should see him eat breakfast—a dozen eggs at least and half a loaf of bread, a big slab of pork with the fat all crispy and brown."

"Be quiet, Ernst." Johann slapped him on the back. "Can't you see I'm trying to eat." He took a large bite of the roll and chewed pensively.

Pete chuckled. "What I can't figure out is why you two big farm boys didn't apply for exemption as farmers. It's a lot easier than going the CO route. The government don't want you to have a conscience. But they kind of like farmers."

Johann popped the last piece of crust into his mouth. "We were stupid, yah. Everyone else applied as farmers, but we didn't know. We just said, we're Mennonites, we don't have to fight."

Ernst nodded. "When the bishop comes with the letter to say we are Mennonites, we are going to tell the tribunal we are also farmers." He chuckled. "And then we can take Johann home and feed him up proper, two turkeys, three ducks and half a beef."

Johann shook his head. David was sure he could hear the big man salivating.

"Way I see it," Pete said, "there's two groups of us—one that'll get the exemption and be sent home to a good meal. That'll be our farm boys, John Henry and the two Friends. They can prove their membership in a 'recognized pacifist organization.' Then there's the rest of us—who are all in a big pickle, garlic, dill and vinegar."

Henri nodded grimly. "Two years hard labor."

"I heard they're going to send some COs over to France—to make an example of them." Bob dipped a small metal cup into the pail of water and drank in big gulps.

David was confused. "Why would they do that? It doesn't make sense to send men to France who refuse to fight."

George shook his head. "It does if you want a spectacle. Look, once you're conscripted, the army figures that you're a soldier. Here in Canada, if you're a soldier and you refuse to fight, the worst they can do is throw you in jail or a labor camp. But if you're a soldier in France and you refuse to fight, they can shoot you."

David was glad he'd given away his roll. He hadn't thought a lot about consequences when he'd made his decision. He'd been too busy thinking about his family, how embarrassed they'd be to have a shirker as one of them. Even at the hearing, he hadn't thought much about what would happen to him if they turned down his exemption.

Now he knew. If they turned him down, he'd be a soldier. And he could either go to France willingly or to jail or, if Bob was right, risk death by firing squad.

He remembered the look of disappointment and pain in his mother's eyes when Mr. Cazen had come to fetch him. Katie wouldn't say Goodbye to him, so he'd left a note and the L.M. Montgomery book he'd been saving for her birthday on the floor outside her bedroom door. How much worse for them would it be if he went to jail or was shot as a deserter? Was it fair for him to put them through this agony for what he believed? Or worse, was it really death that scared him? And was he just a coward after all?

The clink of metal against metal brought his attention to the present. Simon took a drink and passed the cup. "They did that to some English COs, took them to France. Put them through a court martial, threatened them with a firing squad. But in the end, they just took them back to England and threw them in Wandsworth prison. I'm not going to lose sleep worrying about it."

"Sure," Henri quipped, "you don't lose sleep for nothing. Snoring like a steam train all night long."

Simon laughed. "Just poke me in the coal car if I get too loud."

"Candle's getting low," Pete said. "We should turn in."

The men took turns using the toilet at the end of the room. They each carried the candle with them to avoid knocking over the tin can and so they could see to sprinkle the disinfectant over their contribution. David wasn't sure how they'd have managed if Simon and Thomas hadn't smuggled the candles past Sergeant Wilkes in hymn books with the inside of the pages cut out.

David took off his jacket and bundled it into a pillow. Sleeping close together with a few blankets over top of them, David was surprised how warm he felt. His sleep was nevertheless fitful. Dreams he couldn't remember woke him in the blackness. He stared into nothing. The smell of the unwashed bodies around him and the pail at the end of the room kept him awake. David was quite convinced by the end of the night that Simon was not the only one who snored.

At some point, though, he fell into a deep sleep because Pete had to shake him when the sergeant came to the door.

"Hey," Pete said, "are you Ross?"

David snapped awake, trying to remember where he was.

"Ross!" A loud angry voice yelled from the open door-way. It took a minute for David to make sense of what was happening.

"Yeah," he said to Pete, "that's me." He struggled back into his jacket and searched for Pete's face in the darkness. "Thanks. See you later, I guess."

"Good luck, David."

As he walked toward the light, he heard another voice but wasn't sure who it was. "God be with you."

David quick-marched upstairs and was shown into a small room with a table and two chairs. Mr. Cazen waited for him.

"Good heavens, son, where did you spend the night?"

Conscious of how he must look, David tried to smooth his matted hair. He told Mr. Cazen about the room in the basement.

Small glints appeared in the solicitor's eyes and he set his jaw forward. "Excuse me, David, I'll be right back."

David turned his attention to his jacket. A brownish-red stain, five inches across, now occupied the lower left front of the jacket. But David didn't know what it was, or how it had gotten there. He sniffed at it tentatively. It smelled musty, like the rest of him, and it was still wet when he touched it.

The door opened and Mr. Cazen breezed in, a triumphant look on his face. "Well, they won't put you in there again."

"Why not?"

The lawyer sat down and put both his hands flat on the table. He looked hard into David's eyes. "My dear young sir, being a Ross is worth something, you know."

"Oh." It was only that. The Ross name. It wasn't that men whose only crime was not wanting to kill didn't deserve such treatment. It was just the name, the one he was disgracing. Perhaps, after all, there was a better way to do this. If he went to prison, or worse, the Ross name was sure to follow him, making every step of the way more painful for him and for his family.

"Mr. Cazen, I've thought about it. And I've decided to try farming as a profession."

The solicitor nodded and began pulling papers out of his valise. "Your father wanted me to tell you that your cousin Julian has been seriously injured in France."

August 1918
DAVID

AVID POURED COLD WATER into the basin and splashed it onto his face. He paused, then scooped another handful and splashed again. It was early yet; Mr. Tamblyn hadn't come out to the barn.

David was tired. He'd worked quite late on the *Moonlight Sonata*, struggling with the difficult third movement. Afterwards, his sleep had been fitful, like the music.

Or perhaps it was because of what they were going to do today. Going downstairs, he checked in on the hog alone in the pen, the one they were going to butcher.

It was restless too, getting up and pacing around as David approached. As it looked at him with its beady little eyes, David wondered if it was possible for a pig to have a sense of foreboding. Probably not. It was more likely hungry.

Mr. Tamblyn had said they weren't to feed it after yesterday morning; the meat would be tastier if the animal had an empty stomach. Johnny had said the intestines would be easier to clean, then glanced at David to gauge his reaction.

The pig snuffled his near-empty water dish.

"Well, old fella," David said, "slop's out, but I can get you some water at least."

Mr. Tamblyn had said the more water the pig drank, the easier their job would be. David couldn't imagine why this was so but he didn't want to ask. The less he knew about what they were going to do, the less he'd stew about it.

The hog lapped at the fresh water as soon as David filled the dish. David watched and dreaded the day ahead. But Mr. Tamblyn was a kind man. Surely he would find a compassionate way—if there was one—to dispatch the hog. He grabbed a clean bucket and headed down the stalls to start the milking.

He was on cow number four when Hattie and Mr. Tamblyn came into the barn.

"Hattie's going to help us milk so we can get an early start on Bacon and Eggs over there." The large man chuckled, handed a pail to Hattie and left the barn.

"Where should I start?" Hattie asked him.

"I've done The Somme, Lens and Ypres. You could start with Picardy."

Hattie gave him a blank stare. "Which one did you say?"

Her soft brown hair fell in curls to the top of her crisp pinafore. David winced as he felt something twinge beneath his rib cage. "I'm sorry," he said. "I thought you knew their names."

"So did I," she muttered.

Pulling the stool and bucket clear of the cow he was milking, David crossed the barn and stroked a soft brown Jersey. "This is Picardy, my favorite, I think."

He smiled at Hattie, who continued to look at him blankly. "Picardy," she murmured.

"Memorizing cows is more difficult than memorizing music. I completely understand your difficulty. In the stalls, it was easy." He pointed to each animal in turn. "The Somme, Lens, Ypres, Mons, Picardy, Flanders, Gallipoli, Kitchener, Anzac, Princess Patricia, Arras, Haig, Currie and Vimy. But I found once they were out in the pasture, I had no idea which one was which.

"So I had to watch them each carefully and stamp their identity into my head, along with their name. Picardy here was the easiest. She has a rosy glow to her hide, like a Rose of Picardy."

Hattie snorted with laughter.

"No, really," he protested, "I know it sounds odd, but it works."

"Yes, oh yes," she gasped, hugging the shiny pail to her chest. "Except," she sputtered, "which end did you memorize for General Haig?"

David had never seen anyone laugh that hard. She dropped the pail and bent over, holding her abdomen. Eventually, she sank to the barn floor and lay curled up in a ball before she wore herself out. He offered her a hand to get up.

"I wasn't that funny, was I?" he asked in earnest.

"Don't get me started again." Her voice was sharp but her eyes smiled at him—for the first time, he thought.

Hattie sighed. "Who was it told you the cows were named like that?"

Slowly, David began to understand what Hattie thought was so funny. "Johnny." He retrieved the shiny milk pail and handed it to her. "I take it those aren't their real names."

She shook her head.

David couldn't help chuckling, although he suspected Johnny's motive was not humor. "I thought one of them was called something else—Queen of Egypt, you said—but he said your dad sold that one to the neighbors."

"The little blister. I ought to box his ears." She paused and looked at him quizzically. "You must have thought we were stand-to patriots, even naming our cows after famous battles and generals and all?"

David picked up his own pail and stool and resumed milking Mons, or whatever her name was. "Well, I didn't really think it was odd—what with your brother in France."

Hattie set the other milk stool beside Picardy and pulled firmly on the teats, the first streams of milk dancing noisily across the bottom of the shiny pail.

"We're not real patriots, you know, not like some. Will only went, I'm sure, to have a fine adventure. He and all his pals signed up straight away as soon as they were eighteen. Mima didn't want him to go. She's worried herself sick the whole time. And Dada, he'd just as soon Will was here. He thinks farmwork is just as important to the war effort as fighting. And it's been hard to find good help. Oh. Present company excluded, of course."

David patted the cow and carried the full bucket to the door. Twisting the lid off one of the milk cans, he poured the milk into it, the fresh warm smell wafting back at him. "I'm just a city boy with soft hands. Your father's had to teach me every little thing."

"Everybody's got to learn sometime."

"I have to admit," he said, positioning his pail under the udder of the cow known as Flanders, "feeling a bit of a fool spending all that time memorizing the names of cows." He laughed. "They're like spies, I only know them by their code names."

"God, don't talk about spies around Johnny. He's agitated enough as it is—in thick as thieves with that Hamilton bunch—Thatch and Stook are plain bloodthirsty. Kill the Hun and all that."

Huns. Leaning his face against the warm hide of the cow, David lost himself in the rhythm of the milking. Huns. Mr. Liedermann was locked up somewhere in a camp for enemy aliens. His wife had been so hopeful in her last letter that the authorities would let him come home soon.

"Our dear David," she had written, "by now they must know Franz would never hurt a fly and surely he will come home soon. I worry so about his aches and pains. He never does look after himself."

David prayed she was right that they would release Mr. Liedermann soon, but he didn't hold out much hope. Most people thought about the Huns like Johnny did, or like Cousin Julian. Even his sister Katie, in her last letter, had written:

"Julian has been given a big medal for killing all those Huns on his last raid. Mother and Father are so proud, and Uncle Preston has sailed to London to visit him in hospital and see the medal.

"I'm not supposed to know, but Julian has lost his leg. I couldn't help it, I cried all night when I heard. To think he will never be so handsome in his riding boots or dance with Mother ever again. It's wrong of me I know, but I couldn't help wondering if Julian wouldn't rather have his leg than a medal.

"I don't care what Uncle Preston says, you did the right thing because when the war is over, you will come back safe and in one piece, as May says."

The barn door banged and David looked down to see his pail full again. Mr. Tamblyn opened the milk can and held the lid as David filled it with Flanders' milk. It was a powerful thing, he thought, how the milk was replenished every half day without fail. Last night's milking was sitting in cans in the cool water of the river, keeping it fresh until they could take it to the railway for pickup by the dairy.

"Johnny's hitched the team to the wagon. You can take the milk to the train as soon as you're done. And we'll start on the hog right after breakfast."

THE KILLING PLACE was ready. Johnny and David had dragged an old sleigh platform next to a small pen just outside the barn. A rain barrel leaned against the platform at an angle, the bottom blocked so it wouldn't tip. It was half full now with hot water for the scalding.

David laid the saw on the sleigh platform with the other tools. Johnny watched closely as his dad sharpened a six-inch butcher knife, sliding the blade against the steel, first on one side then on the other. Testing the sharpness with the pad of his thumb, Mr. Tamblyn was calm and relaxed.

Johnny's eyes glinted like the sunshine off the blade. "Can I stick him, Da?"

Mr. Tamblyn drew the knife against the steel. "Nope."

Johnny clenched his jaw and balled his fists. "How come you won't let me? I'm old enough. You let Will stick a pig when he was fourteen."

"Will never wanted to as bad as you." David turned. Hattie was steaming toward them with two more buckets of near-boiling water. "Just two more trips," she said, as David relieved her of one pail, lifting it and pouring the water into the barrel. Her eyes were bright and her face pink and covered with a fine layer of condensation.

Mr. Tamblyn laid the knife and the steel on the platform. "Johnny, you help your sister with the water. David and I will bring out that hog."

Inside the barn, the farmer stroked the pig with his hand. "Nice and easy," he said. "We want him to think he's just going for a quiet little stroll." Slipping a rope around the animal's neck, he motioned for David to open the pen. "Let's go, old fella. Just an easy walk." He nodded. "Bring the water dish."

The hog ambled peaceably outside and Mr. Tamblyn coaxed him into the little pen beside the platform. "Give him another drink."

David put the dish where the pig could lap. But he could no longer control his curiosity. "Why give him water now?"

"To keep him cool mostly. That's why you want him calm too—if the pig's excited, his body gets hot and the meat sours."

"Oh." David was a little disappointed. He'd hoped it was a kindness to the doomed animal.

Hattie and Johnny brought more pails. Hattie rolled up one sleeve and dipped her elbow into the water in the barrel. "It's ready," she announced, turning toward the house.

"Aren't you going to stay for the fun?" Johnny yelled after her.

She turned and made a face at her brother. Then, putting her fingers in her ears, she ran for the house.

Mr. Tamblyn picked up the butcher knife from the platform. "Okay, boys, this is how we do this. David, you position yourself along his front quarters. I'll be on the same side, by his hind legs. When I give the word, grab his far trotter and pull as hard and fast as you can. Once he's flipped onto his back, David, you grab both front trotters and Johnny, you hold on to the hind ones. That'll leave me free to stick him fast and clean."

Hours later, the smell of the pig filled the small shed they'd used for scraping and butchering. It seemed to David that the smell saturated everything it touched: his clothes, his hair, his skin. His tongue felt thick and sticky in his mouth, and he was sure if he scraped it, his finger would be coated in lard.

Worried he would puke again, he excused himself and stepped outside to breathe some fresh air. The last sunlight of the day lingered in the tops of the trees, casting shadows underneath. Not sure at first what he was hearing, David peered down the driveway until he made out the shape of a wagon.

He stepped back and opened the shed door. "Mr. Tamblyn, there's a wagon coming up the drive."

"It must be Eli. Johnny, you grab that pail of blood, I'll bring the head. David, step up to the house and ask Hattie to wrap a loaf and some butter for Eli, will you?"

David nodded and walked up to the house, letting himself into the kitchen. Hattie stood over a large bowl, beating vigorously with a wooden spoon. He cleared his throat, and she looked up, her cheeks pink with exertion, a small dusting of flour in her hair.

"Ha, stupid, I know, but I wanted to bake something—to get the smell of the rendering out of the house."

David nodded. "Your father wants you to pack up bread and butter for Eli."

"Ah!" Her face flashed in anger, and she jammed the spoon into the batter. "That simpering Eli. Dada's paid him back a thousand times for helping out last spring, before you came. Wouldn't take a decent man's wage, oh no. He wants Dada to be beholden to him—forever. Here, Eli, have this fat chicken and this dozen eggs and this jug of cream and on and on.

"Now he's getting the pig's head and the blood but that's not enough. Some of Hattie's bread, Eli? Some of Hattie's butter, Eli?" She picked up a jar from the counter. "Might as well have some of Hattie's jam to go with it."

Slamming the jar on the counter so hard David was afraid it would break, she dusted her hands furiously on her apron and stormed into the pantry.

When she came back, she added the jam to a cloth sack and laid it in his arms. "Sorry," she said, "I just hate it. He takes advantage, you know?"

David nodded and smiled. "That's okay." And it was.

He couldn't tell her, but when her eyes flashed and she ranted and raved, she was like the crescendo of a Beethoven symphony.

David joined the men as they stood by a wagon hitched to two heavy draft horses.

"Eli, this is our new hand, David Ross. David—Eli Hough, our neighbor down the line."

The small man smiled broadly and gripped David's hand hard. It seemed natural enough, but David got the odd sensation that he was being carefully studied.

David handed him the sack.

"Now, Mr. Tamblyn," Eli said, looking inside, "you're too kind to Eli."

His voice was thick and smooth, David thought, like tea with too much honey.

"Not at all, that's just a bit of bread and butter. Know how you like it." Mr. Tamblyn shoved his hands into the pockets of his overalls.

Johnny covered his mouth with his hand as Eli set the sack on the wagon seat as if it were a babe in arms. "Eli's obliged, Mr. Tamblyn. Mama makes the head cheese just like in the old country. "

"Not at all, Eli. We're obliged to you. Adeline's not up to making it this year, and I wouldn't want to see the meat go to waste."

Eli sprung onto the wagon box and grabbed the reins. With a click of his tongue and a drawled "Good evening," he was gone.

"Good riddance." Johnny spat on the ground and went back to the shed.

David lingered in the yard, postponing smelling and tasting dead pig. Odd, David thought, that Mr. Tamblyn remained too. Usually the older man was the first to resume work after a break. Together, they watched the wagon lumber out of sight. David imagined the hog's head lolling around in the wagon, the pail of blood sloshing about.

"When you bring something into the world, be it beast or man, you feel responsible for it somehow," Mr. Tamblyn said.

David turned, the question on his face.

"Eli." Mr. Tamblyn indicated the route the wagon had taken. "I stopped by their place that day. His mother had asked me to pick up some needles for her on my next trip to town. With her baby due anytime, I figured she needed them sooner rather than later. When I got there, her sister had gone for the doctor and she was all alone—in labor and scared out of her mind."

David tried to imagine what a soft city boy like him would have done and shuddered.

Mr. Tamblyn chuckled. "I was pretty scared myself. I'd brought lots of calves into the world, pups, kittens, but a baby? I tried to calm her down, hoping the doc or a neighbor woman would get there. But they didn't."

His voice dropped to a whisper so low that the soft rustle of the leaves in the distance competed for David's comprehension. "He was blue and cold. I stuck my finger in his mouth, tried to clear the airway. Blew breath in him. Again and again. I thought … "

The older man's voice sank under the rustle of the darkening evening. "But you saved him?" David said.

"I expect it was only a minute or two, but it seemed like forever. It took too long—he's never been quite right—a worry to his mother."

They turned in unspoken agreement to return to the shed. David thought about Hattie, how her vehement hatred of Eli had made her eyes sparkle. Worlds apart, he thought, from the story he'd just heard.

April 1917
DAVID

ON THURSDAY, DAVID stopped at the Liedermanns' apartment on his way home from work at the bank. Ever since her husband had gone to the camp for enemy aliens, Mrs. Liedermann had encouraged David to practice on their piano, knowing he wasn't allowed to use the one at home. Her eyes were red and swollen when she answered the door.

"I've dreaded telling you most of all," she explained. "Some busybody wrote Franz and told him I wasn't eating right. He insisted I sell it." She pointed to the empty space along the wall. "I wrote to him. I said I can't sell your heart and soul. He wrote back and said that pile of wood and ivory is not my heart and soul. We can always get another piano."

She passed a damp handkerchief across her eyes. David didn't know what to say.

"What will you do, David?"

It was so like her, David thought, to be more worried about how he was going to practice than about how she was going to live from day to day. "Don't worry about me. I'll figure it out." He smiled. "Do you remember when I learned *Fur Elise*?"

She nodded.

"David, he would say, sometimes it takes 300 times to learn a bar of music. The first 299 times you think, I will never learn this, it is too hard."

Her eyes danced behind water. "Then, click, you figure it out and you never forget it."

She touched his arm with concern as he turned to leave.

"Don't worry about me," he said. "I'll figure it out."

The next day, he arranged for a portion of his salary to be placed in an account in Mrs. Liedermann's name. He would give the passbook to Mrs. Campbell. She would know how to convince Mrs. Liedermann to accept an anonymous gift.

Then he went to see his supervisor at the bank, Mr. McCaffery, and volunteered to take over Mr. Nedry's job. Nedry, a middle-aged man with a weak heart, had collapsed under the strain of an essential position and the long hours it entailed. If Mr. McCaffery was surprised, he did not acknowledge it. But David's father called him into his study as soon as he got home.

"Please explain," he said, in the tone David knew he reserved for meetings with junior clerks, "why you would volunteer for an essential position?"

David's life had been the topic for discussion in previous talks. But this time David felt differently. He had made the first move, and for once he didn't feel as if his father held the upper hand of morality and righteousness. His family, especially his father and Uncle Preston, wanted him to enlist, but even if he thought he was prepared to kill another person, it was their determined patriotism that was responsible for all the pain the Liedermanns suffered. His father loved numbers so he would stick to facts.

"It has to be done and I can do it."

His father snorted. "McCaffery is thrilled, no doubt, to get a husky young lad to work like a draft horse. What about your duty? Do you have no feeling for that at all?"

David stood up. There was no point in dragging this out. "Whatever my duty is, sir, it's mine." He paused. "Have him fire me if you don't like it."

The shock on his father's face melted into anger. David left and went into the parlor; he'd wanted to do this for so long. Pull the doily from the piano, lift the lid and sit on the stool. He played Fur Elise, his eyes closed, hitting every note just so—the way it happens only once in many times through. The last note hung, then faded into silence.

"What the hell?"

"Shush."

His parents stood in the doorway, his mother with two fingers on her husband's lips.

"Beethoven," she whispered.

He turned on his heel, leaving her hand a suspended gesture, one relinquished tear on her cheek.

September 1918
HATTIE

Dear Hattie,

Sorry I haven't written in a while, but I've been a bit busy with Jerry. We're just back to billets from the line. I won't mention the place so the censor won't feel it necessary to blacken my good letter.

Except to say that it was a veritable breeding ground for large rodents. We didn't see too much of them in the daytime, but at night they ran like mad up and down our bunks. Right across our faces, bold as anything.

One of our new fellows from the last draft is from Alberta—says he's shot lots of muskrats and gophers (some kind of small prairie dog) but for real sport, he says, nothing beats lying in your bunk in a trench on the line and shooting rats with your .303 service rifle.

The rats here are daring. We are more like the squirrels we used to watch for hours in dear old Sherwood Forest. Every move we make is quick and deliberate. And when we're not moving, we freeze—completely still, hoping our mud-soaked uniforms will be enough to blend us into the rest of the scenery.

So now that we're off the line and pay parade is tomorrow, me and some of the other squirrels are planning a little bombardment of our own at the

local estaminet—a little vin blanc and vin rouge as
a change from Jerry's pineapple and sausage.
  Write again soon.

                                          Love to all,
                                              Will

The letter worried Hattie. Will had never wanted to
drink booze before. When Frankie Hamilton stole a
bottle of his pa's whiskey on his sixteenth birthday and
got Thatch and Stook all liquored up for a joke, Will said
he only took one swallow and didn't like it much. Now
he was planning to get drunk, and it scared her. This was
one letter she didn't want to savor. In fact, she couldn't
get it into the box fast enough. She had no worries about
the woods on such a warm and brilliant afternoon.

Not wanting to seem in a hurry, Hattie ambled, stor-
ing like a squirrel the lushness of late summer against
the coming winter. A robin, fat from a summer's worth
of worms, landed on a branch in front of her. Preening
his breast feathers with a quick duck of his head, he flew
off again.

She loved this time of year when everything felt complete. The corn was full, the tomatoes were scarlet and
plump. The trees had as many leaves as they would get,
but none had started to turn color. Hattie wasn't the only
one to feel this way—Mrs. Nelson had written to say that
"the prospect of the harvest was drawing her home." Even
the war news was good. The German advance had been
broken, and the Allies were starting to push back. Maybe
it would end in time for Will to come home safe.

The thick foliage around the clearing filtered out the brightest light so at first she didn't understand what she saw. As she got closer to the rock, she saw that the sod covering her hiding place had been tossed aside and the hole underneath was black and empty. Frantic, Hattie knelt on the ground and dug through the soft dirt. But it was no use; the box was gone.

That potlicker of a Johnny, she'd wring his neck. It had to be him. He was the only one who knew about the box, knew what was in it. He'd snooped in it before and was probably cross that she'd hidden it on him. She'd box his ears, she'd ...

"Looking for something, Miss Hattie?" A man stood in the shadows at the edge of the clearing. But she didn't need to see him to know who he was. The whingeing in his voice was unmistakable.

"What are you doing sneaking around our property, Eli?"

Stepping out of the shadows, he sauntered toward her, twisting the cuff of his red plaid shirt. "You shouldn't talk that way to me, Miss Hattie. Not at all. It just might be that Eli can help you find what you lost."

She stood up quickly. He sounded too smug, as if he really did know where the box was. But how could he?

"Eli, if you know where it is, you'd better tell me."

Eli shook his head slowly. "Uh-uh, now, Miss Hattie. I think maybe you need to be nice to Eli before he helps you find it."

Nice? After everything Dada had given that smarmy little mooch? She'd be dammed if she'd be nice to Eli Hough. "You can go home now, Eli. I don't need your help."

She could see him smile, even in the filtered light, his small stubs of teeth almost glowing in the dark. "I think you're wrong, Miss Hattie. I don't think you'll be finding anything without Eli." He walked toward her until he reached the edge of the large rock. "No, ma'am, you won't be finding any box, or any letters, without being nice to Eli."

Hattie swallowed hard. If he'd seen Johnny dig it up, he'd know it was a box. But how did he know there were letters inside? She didn't even want to imagine it. "Listen to me, you little maggot," she hissed, "you get off our land and stay off. I don't need you for anything."

The smile melted off his face and he pointed a grubby finger at her. "That's where you're wrong, Miss High and Mighty. You need Eli. Unless you want Mrs. Hamilton to get Frankie's dice back along with the letter that tells her how he really died?"

Hattie wanted to step back out of his reach, but her legs wouldn't obey her. She felt a shudder start at the base of her spine, rising until it shook her shoulders. Eli had the letters.

Hattie remembered how many times she'd been spooked in the woods, sure that someone was following her. There was a queer feeling in her chest and a sinking thought in her head that she'd seriously underestimated him.

"You reckoned Eli was stupid. Now you know different. You have to be nice to Eli." He smiled, as if to himself, and pushed one sleeve of his oversized shirt past the elbow, flexing the sinewy muscle of his forearm.

Hattie could hardly stand to look at his smirk. He was thoroughly enjoying his new-found power. "What do you want?" she asked, anxious to get this over with.

"What would Eli like?"

She should have known he'd want to drag this out—just like he'd milked every possible favor out of Dada for his little bit of spring planting. Just then, someone hollered in the yard and gave her an excuse. "Dada wants me, Eli. Make it quick."

His face slumped. "Fifty bucks."

Hattie was stunned. "Fifty dollars?"

"That's right, you want your box back, you bring fifty dollars to Nelson's barn on Saturday night. And come alone, or you won't get nothing."

"I don't have money like that."

"Hattie!" It was David Ross.

"You can get it—if you want it bad enough." He pulled a crumpled envelope from his pocket and slapped it onto the rock. Then he turned, headed for the safety of the woods to the west.

Now that her legs could move, she couldn't run fast enough. Heedless of the branches whipping across her face and the roots on the path, she raced through the woods. Smack into David.

"Hattie! What are you doing? Are you all right?" She could barely look him in the eye. Everything was at risk right now, and it was all her fault.

"Hattie, what happened? You're pale and you're trembling." His long arms went around her shoulders and for a minute, she leaned against him. He felt warm and strong and just for a minute, she felt safe.

Then gently, she pushed him away. "It's no use," she said. "It won't help."

"Tell me." His voice was so soft and coaxing that she blurted it out, though she hadn't meant to tell anyone.

"It's Eli Hough. He's stolen something and he wants fifty dollars to give it back."

"Tell your father, Hattie."

Hattie wished she could tell her father, but it was impossibly complicated. First, Dada was soft on Eli. Then there was the trouble Eli caused between her parents. But the worst thing was that if her father got the box back, he might see the letters that Hattie had kept to herself—like the one about Frankie Hamilton. "No," she pleaded, "Dada mustn't know. Promise me you won't tell him."

She looked up at him, trying to read his countenance. His fingers grazed her cheek. "I promise," he whispered. "What did he take?"

It was too much. She choked and sobbed once. "Will's letters." Breaking away, she stumbled toward the house and the silence of the kitchen.

THE TELEGRAM CAME on Wednesday morning when Mima was alone in the house. The men were in the north field and Hattie was pulling the last of Mrs. Nelson's carrots. When she came back to start dinner, she found her mother collapsed on the kitchen floor, the yellow paper still clutched in her hand.

Hattie sank to the floor beside Mima and pried the paper loose. Dear God, no, please God, not Will.

"We regret to inform you, Pte. William J. Tamblyn wounded in action August 25, transferred to 23rd Field Hospital, France."

Hattie could only feel numb. She wanted to think. Thank God he's only wounded, but she couldn't. Jimmy and Tom had only been missing in action in their first telegrams. Missing meant dead. Wounded could mean anything.

Her heart pounding in her ears and her hands shaking, Hattie checked her mother's pulse. Hattie put a cushion under Mima's head. There was a lump the size of an egg, but she was breathing okay. Then she forced her rubbery legs to run through the stubble to the north field. The soft dirt slipped out from under her feet at each step and it seemed to take forever to reach the spot where the three men were stooking hay.

Her father dropped his stook and ran to meet her, gripping her shoulders, meeting her gaze. "Will?" he asked.

"He's wounded. Mima must've fainted when she read the telegram. She's breathing, but she's still passed out."

Letting go of her, he waved to the others with one arm. "Come quick, boys." The three of them ran with long strides, leaving Hattie to trudge along after them holding the stitch in her side. By the time she got to the house, Johnny had gone for the doctor, and David and her father had moved Mima upstairs to her bed.

EVERY MINUTE OF the rest of the week dragged like a fork across a blister. Mima woke up, but Doc Watson said her heart was weak and that no one should trouble her. So Hattie tiptoed up and down stairs with tea and bread and butter and soup, in between cooking meals and doing chores.

She couldn't stop herself from wondering if it wasn't her fault, that the box of letters was what had kept Will safe 'til now. She knew it didn't make sense, but all the same, every time she'd put a letter in the box, she'd said a prayer that Will'd be safe. And after all this time, it was hard to think that it hadn't made a difference.

But eventually it was Saturday morning, and Hattie was no closer to having fifty dollars. Johnny and David Ross drove out of the yard with the wagon as Hattie rolled biscuits on the table. She'd have breakfast ready by the time Johnny got back. He'd be grouchy, of course, like he was every Saturday. He begrudged anything that had to do with David Ross. And going to the railway on Saturdays to bring back the team so that "the conchie" could take the train into the city to sign in at the depot was the thing he hated worst of all.

But today, Hattie scarcely cared when Johnny stomped into the kitchen. She had to decide what to do about Eli today; she had to get Will's letters back.

She would go to Nelson's barn, she decided, and take her grandmother's cameo choker. It was the only thing of value she owned. And she'd bake pie, fresh apple pie. Her stomach churned at the thought of baking for Eli Hough, but he was always full of compliments about her cooking and right now, with Will's fate hanging in the balance, she'd do anything to get those letters back.

When the pies were cooled, Hattie put the biggest one in a cloth sack, suspending it carefully from a tree just inside the wood. If Dada asked, she would tell him she was going to check Mrs. Nelson's house. But it would look a big lie if she took an apple pie with her. With the cameo safe in her pocket, Hattie was ready.

Supper was a quiet meal. Johnny was sullen, her father tired, and David Ross hadn't gotten back yet from the depot. Hattie was sure they didn't even notice when she slipped out and headed for the woods.

"Hattie." The voice was quiet, but she was so startled that she nearly dropped the pie.

"David Ross, scare a body to death!" She held the sack tightly with both hands.

"I'm sorry. It's just... I wanted to catch you alone." He held his hand out. She reached hers toward it. It wasn't what she thought. He pressed some paper into her hand.

Bringing it close to her face, she couldn't believe what it was. "Fifty dollars?"

"I still think," he said, "that you should tell your father. But if this will help, it's yours, and don't worry about paying it back."

Hattie folded the money and put it in her apron as a last resort. She didn't want to be beholden to David—certainly not for fifty dollars. But if it was the only way to get Will's letters, she'd use it.

The long walk through the woods gave her the chance to speculate about the man who'd given her the fifty dollars and about the man who wanted it. Where did David Ross get the money in the first place? He didn't seem the sort to steal, but Hattie knew Dada wasn't paying him wages. His only possessions, besides his shaving kit and his work clothes, were his cloth piano and his sheet music. And though he went to the depot every Saturday in town, Hattie had never seen him bring anything back. Really, she knew very little about him and what his life was like

before he'd come to the farm. He'd told them he lived in
Toronto with his parents and little sister, but in all the
months he'd been here, the only other thing Hattie had
learned was that his sister loved the Anne books.

She decided it didn't matter where the money had come
from. She had no reason to think it was ill-gotten, so if it
was—it was David's problem, not hers.

The thing that bothered her most was why, if Eli really
wanted fifty dollars, would he try to blackmail her for it?
He was cunning enough to know she didn't have money,
which gave her the sinking feeling that it was something
else he wanted.

Hattie propped open both sides of the barn door so
the moon would light the inside. Retrieving the lantern Mrs. Nelson kept behind the door, Hattie lit it and
looked around. It had been such a long time since cows
occupied the stalls that it didn't smell like a barn, just the
warm dusty scent of old hay and a hint of sweet clover.
The enormous stack of hay in one corner, scattered and
untidy, and an old milking stool were the only clues the
barn had ever been used at all. Pulling the stool into the
open door, Hattie sat down with the lantern and the pie
beside her. She wanted to be prepared for Eli.

The wind picked up, humming through the cracks
between the boards, skittering a piece of dried lilac across
the yard, and still Eli didn't come.

She watched the tree line closely; once or twice she
thought she saw a large shadow move with intent, a shadow
that couldn't be a tree or a raccoon. It'd be just like that little weasel to skulk around in the bushes, Hattie thought.

Taking the lantern, she ventured out into the yard, trying to angle the light into the shadows.

"I'm not waiting all night for you, Eli!"

The shadows softened into the tree line and she received no reply. Until someone grabbed at her wrist from behind.

"Aaahhh!" Hattie flung her wrist as if she was trying to throw off a snake. She whirled around, the lantern hissing and sputtering, her heart pounding. "Don't touch me."

Eli smirked, picking at one dirty fingernail with a small stick. "Eli scared you."

Furious, Hattie didn't want to admit he was right. "Where's my box?"

"It's here." He shrugged.

"Then give it to me."

He shook his finger. "You brought the money? You're not getting the box 'less Eli gets what he wants."

"Show me the box first." She took a couple of steps toward him.

"Maybe Eli don't want to."

The scowl on his face worried Hattie. Maybe she'd have to try something different.

"Sure you do," she said, forcing a smile. "Show me the box and I'll give you the apple pie I made."

His eyes sharpened. "Pie first."

Giving him a wide berth, she walked behind him and picked up the pie. He followed, practically grabbing it from her hands. Unwrapping the muslin, he dug a piece out with his hand and ate it noisily, catching the drips with his tongue. Hattie stepped back. She shuddered and by the time he'd licked his fingers, Hattie's stomach felt queasy.

"The box," she said. "Where's the box?"

"Miss Hattie in a hurry?" he slurred, his tongue thick with the sweet confection.

"You were late."

Bits of apple stuck to blackened teeth as he smiled. "Eli wasn't late. Heard you open the door, saw you light the lamp. Eli's got what you want. You'll wait."

Hattie wanted to squeeze him like a pimple, but she knew it was hopeless. Until she knew where the letters were, she was at his mercy. Setting down the lantern, she folded her arms and pinched the soft flesh underneath until it hurt. If he wanted her to wait, she'd wait. But he'd have to wait too.

She stared straight ahead, pretending not to notice the details as he slurped down another handful of pie. He wiped his hand on his pants and set the pie on the stool. He stood close enough to her that she could smell him, but she stared right through him. Picking up the lantern, he retreated into the very back of the barn and rummaged in the hay stack. Hattie waited.

"Better come on back here."

She said nothing. It was his turn to wait.

"Too proud for your own good, Miss Hattie." He pulled out a letter, opened it slowly and scanned it as if reading it. He scraped at something on the paper with his thumbnail. "Careless," he said, "to spill wax all over his letter."

Hattie's stomach clenched. She knew which one he had—the one Will was so proud of. It had started with Jimmy, writing a letter in a trench somewhere at night, using a stub of a candle. As he bent over his work, a louse fell onto the paper at the same time as his candle dripped and the hot wax fixed its victim to the page.

Hattie had shuddered at the thought of Jimmy crawling with lice, but he'd turned it into a joke—convinced everyone in the company to write a lousy letter home. Now this letter, that Eli was scraping at, was all she had left of Jimmy and his sense of humor. Hattie wanted to drown Eli in wax, but she was sure that the more attention she paid him, the longer it would take to get the letters.

"Want to see it?" Silence. "Still too proud? I think the wax might set this on fire real quick."

He held the paper to the lantern. The edge of the paper started to smoke.

"No! Stop! I'll give you what you want."

His smile in the lamplight was gruesome. "Come on back here, then."

She had no choice.

The box, its lid ajar, was nestled in the haystack. Eli hung the lamp from a nail on the wall behind him. He beckoned to her with one hand.

"What you got for Eli?"

She decided not to try to bargain him down with the cameo. She shoved the money at him and reached for the box. But he stepped in front of her and grabbed her tight around the waist with one arm.

He rustled the money in her face. "Fifty bucks. Now where'd a little girl like you get that?" The smell of rotten honey filled her mouth and she twisted to get away, but his arm held like a clamp. He shoved the money in his pocket and used both arms to back her against a beam. Sliding one hand along her neck, he clenched a fistful of her hair. "How about a kiss?"

Hattie spit in his face, and he slapped her. Her skin stung in the imprint of his hand and her jawbone ached.

"Let her go. Now!"

Eli released his grip and Hattie stumbled toward the open door. It took her a minute to realize what was happening. It was David Ross, and he'd brought Will's quarterstaff. Maple, Will had peeled it when he was ten and carried it everywhere for years after. It was the one weapon Hattie knew how to use.

David leaned toward her. "Are you all right?"

Her face still burned where Eli'd slapped her. "I'm fine," she said, wrenching the quarterstaff from David's hands.

As she felt the smooth wood in her hands, the training came back to her. Advancing on Eli, she swung the stick above her head and spun it in her fingers before assuming the jousting position.

"Hah!" She leaped forward, poking him as hard as she could in the gut.

Stumbling back, surprise on his face, he grabbed the soapstone box, holding it in front of him like a shield. Hattie didn't hesitate. She smashed his fingers with the stick, crushing them against the box.

He cursed, dropping the box and leaning to the right, cradling his injured hand. Hattie thought about his fetid breath on her cheek, his greasy fingers on Will's letters. She wound up, bringing the staff down behind his bent left kneecap. Eli collapsed to the barn floor and before he could get up, Hattie drove the end of the quarterstaff just behind his ear, hard enough for him to lay still.

"Scurvy dog!"

"What?" David laughed.

Hattie relaxed but held the staff firm. "It's what we called the Sheriff's men when we knocked their blocks off."

Eli whinged. David stood behind Hattie and gently rubbed her shoulders. "I think you can let this one go now," he said.

She leaned back slightly, her hatred for Eli pooling into disgust. "Get out of here," she said, releasing him.

She knelt, gathering the strewn letters. The lid of the box had broken in the fall and she couldn't find Frankie's dice in the dim light.

Eli lay for a minute, then struggled to his feet, leaning almost doubled on his right side. Hattie dismissed it as melodrama, but David moved swiftly, pinching Eli's bruised hand, forcing him to drop a wicked looking blade about six inches long.

"Get out," David said. As Eli slunk out of the barn, David picked up the knife and showed it to Hattie. "He must have had it in his boot."

Hattie shuddered. She'd have been in serious trouble if David hadn't showed up when he did. It was only then she remembered that Eli still had the fifty dollars.

"Maggot," she cursed, pulling the letters into her lap.

David squatted beside her, handing her an envelope. "Are they all here, do you think?"

"I don't know." How could she have gotten herself into this mess? Better to get it over with. "He took your money with him." She watched for anger, but he just shrugged.

"That doesn't matter."

"How can you say that? How can fifty dollars mean nothing to you?" She hated the way he always confused her—so calm and gentle and strong that she wanted to touch him and yet not normal, never what she expected. It made her want to lash out at him, and when she did, she hated herself. One by one, she stacked the letters into the empty box, shoving them hard into the corners.

"It just doesn't," he whispered.

"Did you steal it, then?" her voice a runaway horse that any minute would throw her off and escape on its own.

"Hattie!" He grabbed and held her struggling hand between both of his. "I didn't want to tell you because …" His throat seemed to catch and he cleared it. "The army has to pay me, because of the rules, as if I were a soldier. I didn't want it; it didn't seem right. So they put the money in an account. I was going to leave it there. But, to help you get the letters back … It seemed the right thing to do. But I don't want the money."

The warmth of his hands seemed to soak into her skin and she wondered if it would leave an imprint, long weeks from now, when he was gone. She hated the drops of water that ran unbidden down her cheeks, deserting soldiers that should be shot for treason. As always, he seemed to sense her feelings and released her hand.

She put the rest of the letters in the box, slowly, deliberately, then picked up the two pieces of the lid. "Can't mend stone." A whisper was all her voice had left.

David found a piece of twine and helped her tie the box together. Hattie doused the lamp and hung it behind the door and they walked across the yard together. Just as they reached the cover of the trees, she remembered Frankie's dice.

"We have to go back."

David put a hand on her shoulder. "Do you hear something?"

"No," she said without listening. "It's important." And she told him about the dice. "I have to get them back."

But he held her back when she tried to move past him. "Listen."

She heard it now—horses' hooves and human voices, in the distance, but moving closer.

David handed her the quarterstaff. "You go. I'll wait until they've gone, then I'll get the dice."

"If it's those Hamilton twins, don't mess with them. They'd kill you as soon as look at you."

He moved into the shadows of the tree line. "Don't worry, I won't."

Hattie tied her skirts and trudged home, stopping every few minutes to listen, hoping she would hear David catching up with her. The house was dark and quiet and she waited on the verandah, cradling Will's box for what seemed an eternity, and still David didn't come. Slipping upstairs, she sequestered the box in a drawer underneath a petticoat and was about to go downstairs and back to Nelsons' when she got a strange feeling that something was wrong. Here.

Standing in the hallway in her stockings, her boots in one hand, the quarterstaff in the other, she couldn't hear a thing. That was the problem. The door to the boys' room was open, and it was as silent as a tomb. Most nights she could hear Johnny from her room with both doors closed. She tapped his bedclothes with the staff. He was gone.

September 1918
DAVID

*D*AVID LISTENED FOR THE rustle of Hattie's movements through the brush against the noise of the approaching riders. If there was trouble coming, he wanted her to be well out of it. Although he didn't really need to protect her. The way she'd whaled at Eli with the stick had unsettled him. But like her eyes when she was all fired up, it had excited him too.

She had every right to be angry, he thought, given what Eli had done. But the raw power she unleashed was both brilliant and startling. It made him want to hold her close to him, press his lips on hers. But he also wondered where she would have stopped with Eli if he had not been there.

Whooping and hollering, five or six riders galloped into the yard. A couple of them carried lanterns and one was dragging something behind on a rope. Whoever they were, they weren't likely to follow Hattie, and David started to move back into the shadows to wait it out. Two things made him stop. One of the riders brought his lantern close to his face—it was Johnny Tamblyn. Then the rider with the rope dismounted. He kicked hard at what he'd been dragging.

"Uhhhh." The sound was low, but unmistakably human.

"Guess we have to take you for another ride."

"Owww!" With another kick, the groan became a howl, piteous, like a terrified dog.

"You'll kill him, Thatch," one of the others warned.

"So?" the rider challenged. "One less dirty foreigner on the line. I say we drop him right by his mama's gate."

David felt in his boot for Eli's knife and ran his thumb across the blade. Sharp. But could he get there in time?

David was on the move as Thatch swung into his saddle.

"Lead us out, Billy," Thatch hollered, and one of the lantern-bearers headed for the main road. The others followed Billy while Thatch turned his horse around.

David got one hand on the rope as the horse started to move. He ran with it, sawing at the rope with his free hand until it snapped. Kneeling in the dust, he cut away the rope that bound the man's wrists, before he turned him over. Dirt caked with blood but David recognized the smell of rotten honey. Eli.

He shook him. "Can you get up? Quick! They're coming back."

David couldn't see them, but it sounded as if the horses had wheeled partway down the lane and were thundering back toward them. Eli cried out in pain as David pulled him to his feet. His left arm looked broken. David pressed the knife into his other hand.

"Go! Get out of here." He pointed at the woods behind the barn, where the horses couldn't easily follow.

Eli staggered, glaring at him with fear and suspicion. But as the riders bore down on their position, Eli stumbled toward the brush, each step quicker than the last. David faced the riders.

"Who the hell are you?" Thatch demanded, swinging from his horse. "Spoiling our fun."

Thatch pushed him hard in the chest. David took one step back but stayed there. The longer he could keep them interested in him, the better chance Eli had to get away.

"It's Tamblyn's conchie. Isn't it, Johnny?" one of them hollered.

Johnny's face in the lantern light was grim as everyone waited for his answer. He nodded at Thatch.

"That so?" Thatch shoved him again. "Maybe you'd like to take a little ride with us? Would ya?"

For a second, David saw Julian with his clenched hand in the middle of the stream. Then, before he saw it coming, Thatch's fist connected with his jaw and he hit the ground. Strong arms lifted him from behind and the fist connected again.

"Put the boots to the slacker."

Doubled over from a kick to his stomach, he blacked out for a second as it was followed by one to the kidneys.

"Stook, you bring your fence-pounder?"

Something heavy crashed down onto his knee. Crack. The pain shot up into his thigh and burned.

"Bring me a rope. We'll string the coward up from the loft."

They dragged him by the arms, propping him against the barn door.

"Here, Johnny. He's your conchie—you tighten the noose."

Thatch shoved the twisted rope into Johnny's hands and forced him onto his knees beside David. Afterwards, David thought it might have been the lantern light, but looking into Johnny's face as he turned the frayed hemp in his fingers, David realized that her younger brother had Hattie's eyes. Later it embarrassed him. But at the time, despite the burning in his leg and his jaw, and the fear of what was going to happen, the thought of Hattie made him smile.

Johnny winced and he flung the rope aside. "No, I won't," he said.

"Then I'll do it for ya." It was Thatch.

"No." Johnny stood up. "You're not killing him."

Thatch pushed him. "Who's going to stop me? A boy in breeches?"

"Yeah." It was a voice David didn't recognize.

"Stay out of this, Billy." Thatch said as a small boy with a lantern moved to stand beside Johnny.

"Take the boys and go home, Thatch." Billy's voice was soft and high, but he spoke as if he knew he'd get his way.

"You little whelp." Thatch lunged at him, but one of the others held him back.

"Let's go, Thatch. You know what Mam'll say if you hurt one louse on one hair on his head."

Thatch cleared his throat and spat, then turned with the others. Billy and Johnny watched them go.

"Johnny?" Billy nudged his arm.

"Yeah?"

"Why?"

"I don't know. When I took the rope and looked at him, for a second, it was Will. And I couldn't."

"What do we do with him?"

"Leave him here. If he don't show up by morning, Hattie'll come looking for her conchie."

David waited until they left before he dragged himself backwards into the barn. Sifting through straw and chaff as he went, his fingers touched one, then two, small cubes. Leaning against the hay, he wiped the spittle from his cheek with his dusty sleeve and squinted in the darkness at Frankie's dice.

September 1918
HATTIE

*S*HE SWEPT THE BARNYARD with the lantern. Something had caused a dust-up.

"David?"

There was no reply, so she checked the barn. He was sprawled against the hay stack near the back, one leg twisted at an odd angle. Was he dead, or just passed out? She knelt beside him and touched the angry swelling on his left cheek, and he stirred.

"My God, what happened?"

He tried to smile, but winced. "I think my jaw may be broken."

She held up her lamp to get a closer look at his face. An ugly blackness was already spreading underneath the swelling. "We've got to get you to a doctor. Come, I'll help you get up." Setting her lamp aside, she slipped her arm underneath his shoulder and lifted as he tried to get to his feet.

"Ah." A sharp pain flashed across his face and he fell back heavily. Hattie, her arm still entwined in his, sat down hard.

"What's the matter?"

He drew several breaths before he could reply. "I think the leg's broken too."

"Who did this?" she asked.

He gave her a simple version of what had happened.

"Those twins have always been vicious! Why did you do this after I told you?!" She stopped. "Johnny was with them, wasn't he?"

"No."

Thank God for small mercies. David leaned back, taking short labored breaths. She wrapped her shawl about his shoulders and brushed a hank of chestnut hair away from his damp forehead. How would she ever get him out of here? She couldn't do it by herself and she didn't want to leave him alone. What if they came back? "Eli Hough's not worth this," she said at last.

David's voice was quiet and low. "They'd have killed him, Hattie." He touched her forearm gently with his right hand. "You wouldn't have wanted that, would you?"

Hattie searched her pounding heart—did she hate that weasel enough to wish him dead? Mostly he just disgusted her. But it wasn't that simple—was he worth David's life? "I don't know," she whispered.

Then, his eyes still closed, he said, "Hattie, why is your brother brave? Because he's willing to kill the enemy or because he's willing to risk his own life for what he believes?"

"Will?" Hattie choked. She thought about the yellow telegram, wondered what would come next. "Because he risked his life," she said softly.

David opened his eyes and turned toward her. His hand slipped around hers, dry and warm.

"Your brother and his friends risk their lives for what they believe. I don't want to be less than them."

"David."

"Shh. Let me finish." He winced, taking painful breaths every few words. "When I became a CO ... everyone thought I was a coward. The neighbors ... the people at the bank ... my own family. My little sister—I could see it in her eyes. Coming home one day ... a lady on Spadina ... dusted me with flour. 'Slacker,' she said. What she meant was 'coward.'

"Was it just my lousy hide I was protecting? I could never really be sure.

"The first time I played Fur Elise ... all the way through ... I knew life was a miracle ... and I never want to give it up ... but I couldn't let them kill him." He coughed and tried to smile. "Weasel or not."

He gripped her fingers tightly, but Hattie couldn't say anything. She squeezed his hand and sat quietly, listening to the sounds of the night: small chirps and the rustling of mice in the hay.

SOMETIME IN THE NIGHT, Hattie covered them both with her shawl and they slept in the hay until Mrs. Nelson found them the next morning.

"Hattie?" Mrs. Nelson shook her shoulder. "What's going on? Your father is so worried."

Hattie roused, barely glancing at her neighbor, her thoughts on David. She stroked his face until he opened his eyes, then turned.

"David's leg is broken. He needs a doctor." Hattie thought Mrs. Nelson looked splendid in her blue traveling suit, complete with gloves and hat, but this wasn't the time to tell her.

Mrs. Nelson looked thoughtful for a minute. "Your father is here looking for you—gave me a lift from the corner. Are you all right?"

Hattie got up and brushed away the straw, snugging her shawl around David. "I'm fine."

Mrs. Nelson nodded briskly. "I'll catch them. Good job that Johnny noticed the barn door."

Hattie met her father's gaze as he entered the barn. "The Hamiltons beat David," she said, glancing at Johnny. "Lucky they didn't kill him."

Mr. Tamblyn knelt, checking David's face and his leg. "The buckboard will shake this apart something fierce."

"No need for that," said Mrs. Nelson, taking off her gloves. "Bring him into the house."

They carried David across the yard and Mrs. Nelson directed them to the four-poster bed in the ground-floor bedroom. David faded in and out of consciousness as Mrs. Nelson directed Hattie in the business of making him clean and comfortable. Hattie sponged his face and neck with warm water and salts.

"He'll be all right, won't he?" she asked Mrs. Nelson.

Her neighbor shook her head. "I hope so. This is a bad job. If they didn't burst something inside, and if the leg can be mended, he's young and strong enough to bounce back."

Hattie didn't want to hear her fears confirmed; she wanted to scream, to cry, to pray. But who would listen? How many lives were being prayed for right now? David woke as Mrs. Nelson went to the kitchen. He grabbed Hattie's hand and pressed it to his heart. No, to his shirt pocket. She looked at him, puzzled. He smiled and nodded a little. Undoing the button on the pocket, she retrieved the two small black dice.

"Oh!" she said.

"It'll be all right now," he whispered.

Mrs. Nelson returned with a sharp knife and, declaring her intention to cut away David's pants, she shooed Hattie onto the porch with her father. Johnny had already taken the wagon to fetch Dr. Miller.

Mr. Tamblyn looked hard at his hands. "I don't suppose you know what he was doing over here at night?"

"No, Dada."

He looked up at her. "What were you doing?"

She'd had some time to think this one through. "I came early to dig potatoes and the barn door was open, so I checked inside."

"Uh-huh." Hattie knew the tone of his voice meant he didn't really believe her. "Do you think Johnny had anything to do with this?"

This time she could answer with the truth. "I thought he might have, but I asked David, and he said Johnny wasn't there."

Her father sighed. What kind of life, Hattie thought, when this was the smallest of mercies?

April 1919
DAVID

AVID WALKED UP the long gravel drive to the gray stone building. His leg had healed well, leaving him with only a slight limp. From time to time, though, he stopped and bent his knee, then stretched it. He wondered how long this habit would stay with him.

In his right pocket, he could feel the bulge of the brown paper bag his sister Katie had thrust at him as he left the house.

"Please," she had said, "tell him they're from me. He always brought me orange drops."

David doubted he would ever have come here, if it hadn't been for Katie.

She had begged him. "Please go and see Julian. Father has visited him with Uncle Preston, but he won't let Mother and me go. He says there's no use, that Julian wouldn't know we were there. Please go. There's got to be some part of him that's left—some part that would remember."

Having disappointed her in so many other ways, David couldn't bring himself to deny her request. He'd retrieved the address from the blotter in his father's office and said he would go.

He pondered the large building with the open windows and cheery flower beds, wondering exactly what he would find inside. He knew his cousin had lost a leg, but that wouldn't have made his father forbid his mother and sister to see Julian. There had to be something more.

Two men in wheelchairs sat on the wide verandah, one heavily bandaged on the lower half of his face. Neither seemed to notice David as he walked across the stone floor and through the French doors into the spacious, sun-filled front hall. Men sat on couches and chairs scattered around the room. Two played a game of chess at a small table in front of an upright piano.

On the far side of the room, beside a sunny window with a wide ledge, was Julian. He was seated in a wheelchair, dressed in a morning suit with one pant leg neatly pinned underneath. He reached out, deposited some small thing on the window ledge.

David would have known him anywhere. He was a bit thinner, of course, but his chiseled handsome profile was the same as ever. Pulling the bag of candy from his pocket, David crossed the room to his cousin's chair before he noticed what was wrong.

Julian's eyes, always so cunning and intelligent, were lifeless. The brilliant blue irises still glistened in the sunlight, but the person behind them was gone.

David could hardly believe it. "Julian?"

"Willows!" A shrill voice pierced the quiet room, and David looked up to see a young nurse steaming toward them. Involuntarily, David took a step backwards. The nurse dusted off Julian's lap and shoulders with sharp, brisk movements, then grabbed the handles of the wheelchair. "Willows!" she yelled again, pushing Julian's chair toward the center of the room. David followed.

"Are you a friend of Captain Ross?" she asked him, her small dark eyes full of veiled suspicion.

"No, I'm his cousin, David."

David was sure it was fear that leapt into her eyes. Knowing his Uncle Preston, he knew why.

A young woman in a striped dress and with untidy red hair scuttled into view. Willows, responding to the call, David thought.

"I've told you before," the nurse told her, "not to park Captain Ross beside the window."

"I'm sorry, ma'am."

The nurse's voice had an edge of icy sweetness. "Get a brush and a dustpan and clean it up."

"Yes, ma'am." The girl bobbed and left as quickly as she'd come.

The nurse smiled at David. "Is this your first time here?"

"Yes, I've, uh, been away." He bent his knee and stretched it. Bad habit. He lifted the small bag. "My sister sent some orange drops. May he have them?"

"Oh, of course." She took the bag, rolled down the top and placed it in Julian's lap. "There you are, Captain Ross, a morsel for your sweet tooth."

"Does he understand what you say?" David asked.

She looked surprised. "I don't know. I think it can't hurt to try," she said.

"Does he ever speak to you?"

"Oh, no, not at all."

David pulled up a small wooden chair. "I'll just sit with him a while, then, if that's all right."

"Of course." She smoothed her uniform and backed away from them.

David touched the bag of candy lightly. "Katie sent the orange drops. She remembers you used to bring them for her. And mother sends her love. They... would come, if they could, if father would let them."

Julian didn't seem to notice David or the candy. He turned his head over his right shoulder, as if he were trying to see something. Maybe it was the window, or the ledge.

David walked over to where his cousin had been when he first arrived. And he saw what Willows had been instructed to clean up. Turning away from it, standing in the full, hot light, he hoped the sun would burn from his mind the image of flies with their wings carefully detached.

When David finally turned to look at his cousin, Julian was no longer twisted toward the window. He sat straight ahead, staring. David walked past the wheelchair on his way out. He stopped, touched one sleeve lightly and said, "Goodbye, Julian." His cousin made no response.

David couldn't get out of there fast enough. But before he could escape through the French doors, he was hailed by one of the chess players.

"Sir, do I know you?" The man looked too old to have been in the war but his hoarse voice and the scarring on his neck identified him as the victim of a gas attack.

"No, I don't think so." David spoke softly.

"Let me look at you. No, maybe not. I took you for one of Captain Ross's mates, coming to see him and all. I was in his company, you know." The man toyed with a chess piece in his hands as he spoke.

"I see. Captain Ross is my cousin." Bend the knee, stretch it.

The old sweat shook his head. "A shame that his mind was affected. A scratch like that leg would never have kept a man like him down." He pointed at David with the chess piece in his hand, a black knight. "Your cousin was the bravest officer I saw in three years, never shirked behind the lines. Volunteered for every trench raid there was."

"Umm." David tried not to think about flies or kittens.

"He saved my life once, he did." The man played his knight on the board.

His partner snorted. "Don't tell that old story again, Dutch."

"Why not? He hasn't heard it before, have you, sir?"

David swallowed hard. "No."

"There, see. We went over on a trench raid in full moonlight, more fool us. Threw our bombs, thought we'd finished Jerry off. We're in their trench looking for souvenirs, when this big Heine comes up behind me. I don't even see him or know he's there until Captain Ross leaps past me with his bayonet drawn.

"In, out and on guard! Just like he was doing it on the practice field with a bag of straw. Then he says, 'Watch your back, Dutch' and pats me on the shoulder. Best damn officer in the regiment if you ask me."

A spasm seemed to catch in Dutch's throat and he coughed violently, hacking uncontrollably for so long that David became alarmed.

"Should I get the nurse?" he asked the other chess player.

The man shook his head. "Nothing anybody can do—the dirty Jerry gas."

Hunched over, Dutch continued to cough for several minutes, the severity of each spasm gradually easing up and the time between each one lengthening. Though he was anxious to leave, David didn't move, not wanting to add insult to the other man's injury. Eventually, the coughing stopped and Dutch sat up, his eyes watery from the exertion.

"Thing … remember," he rasped. He pointed at his friend. "Voice … song."

His mate nodded. "I remember that—before I was wounded. We were wasting away in the service trench, lousy rations, lousy mud, lousy rats, lousy us. It was worse than being on the line, because you had time to think about what might happen to you when the show started. One of the sweats won a squeeze-box in a friendly game and started playing songs to pass the time. He knew 'Roses of Picardy' and when the first chord played, this fine tenor voice sang out. Beautiful it was."

David was nonplussed. "Julian?"

"Captain Ross," the man nodded.

He'd never heard his cousin sing. It wasn't something their families had ever done together. All their lives, David had thought of his cousin as his complete antithesis.

He was dark, Julian was fair; he was shy, Julian was outgoing; he was sensitive, Julian was hard; he ran with endurance, Julian with speed. There had never been a point of contact, a place they both occupied at the same time.

Dutch swallowed hard. "I'd give a lot to hear that again."

Glancing at Julian, David knew there wasn't much hope. But perhaps there was a way. Stepping behind Dutch's chair, he moved the stool closer to the piano. He chorded an introduction and began with the chorus.

"Roses are shining in Picardy, In the hush of the silvery dew..."

When he was finished, the room was hushed like a church.

"That was fine," Dutch said finally.

"Captain Ross'd be proud," his partner added.

The old fears that had threatened to overtake him were gone. All that was left was pity, but these men did not deserve that, he thought. Instead, he focused on why he was here, why he had come. He offered his hand. "I'm glad we met, Dutch. I'll be able to tell my sister that Julian has comrades around him who think very highly of him. My father doesn't want her to see Julian the way he is, but she is very fond of him. He used to bring her candy."

Dutch shook his hand and nodded, making a rumbling sound deep in his throat. David excused himself and strolled down the path, the need for urgency gone. He didn't see Willows dust the window ledge or Julian dig in the small paper bag for an orange drop.

May 1919
HATTIE

MRS. NELSON WAS A woman on a mission. When Hattie arrived to help her pack, there was barely space to walk between the trunks and crates on the parlor floor. China cups, blankets, pots, blouses and photographs were piled on the tables and the floor. "A powerful determined woman," Hattie's father had said, when Mrs. Nelson announced her decision to sell the farm and devote herself to the education of rural women as a traveling speaker for the Women's Institute. Inspecting a brown taffeta skirt, she waved Hattie in.

"Only so much room," she said, folding then rolling the skirt into a tight package. "But I'll need something for Sundays." She tucked it into a small wooden trunk at her feet.

"What can I do?" Hattie asked.

Mrs. Nelson's face softened. Deftly stepping over the debris, she threw one arm around Hattie's shoulders and squeezed.

"This is very kind of you to come and help me with my foolishness when you have all of your own troubles at home."

Hattie didn't know what to say. How could she tell Mrs. Nelson that it didn't matter what kind of work she was asked to do here—it was a relief to be out of the house and away. Away from her mother staring into space over a cup of cold tea, away from Johnny's smoldering moods. Instead of sitting on the verandah in the evening with a piece of whittling, her father retreated to the barn after supper. Last night, Hattie'd found him straightening old nails with a hammer and a pair of pliers. She felt a familiar squeeze on her heart—the one she'd felt more and more since David Ross had gone away.

She hadn't realized until he'd gone that he had begun to fill so much of her life. Now every time she opened a closet or walked into a room, it felt as if something were missing. As thoughts crowded her brain and she had no one to tell, she felt empty inside. It was like when Will had first gone away. Only now it was worse, because there were things she could have told David that she would never have told Will.

Trying to shake away the thoughts that often produced tears, Hattie shook herself loose from Mrs. Nelson's embrace as well. Afraid she'd offended her, Hattie murmured, "Sorry."

But Mrs. Nelson was unruffled. "Quite all right, my dear. Let's get to work—sometimes the monotony of tasks is the only medicine the day provides."

Hattie managed a quick glance and a nod in reply.

"There's a pile of clean rags on the kitchen table. You can use them to wrap the china from the cabinet. All the dishes can go into the wooden crates. My brother-in-law is bringing a wagon next week to take everything back to Kingston."

Hattie loved wrapping the delicate china plates, folding the flannel strips around them as she piled them in the box. She picked up a cup Mrs. Nelson had once served her tea in. Turquoise with gold trim, she had forgotten how light it was, but she remembered how fine she'd felt, holding the tiny handle and slipping her tongue just over the delicate lip of the cup.

"Won't you miss it?" She blurted out.

Mrs. Nelson smiled. "Miss what?"

Hattie held up the cup. "Your things, having your own home."

"I'm not sure. There is a certain comfort in eating off the same plates and sitting in the same chair day after day. It makes you feel like you belong. But maybe that's a bad thing. I don't belong to this house or to these trinkets. I didn't belong to John either, although for a long time that's how it felt. I don't even belong to my memories, although unlike this lot, I'll have to take them with me."

Mrs. Nelson took the Captain's picture from the piano, wrapped it and placed it in the trunk closest to her. "I belong only to myself and to whatever I am called to do. Maybe living without all this will help me to remember that every day, instead of once in a while."

By late afternoon, they'd cleared all the drawers and cupboards of their goods. The crates and trunks were covered, the concealment of their contents making the parlour and the dining room look as empty as a train station.

They sat in the kitchen, drinking from the only cups that had escaped the packing. Hattie stirred two spoonfuls of sugar into the hot liquid and gulped back a scalding mouthful. She thought she understood why Mrs. Nelson wanted to go away, but she was also angry—at Mrs. Nelson, Will, David—everyone who went away while she stayed behind.

"Have you heard from the young man?"

Hattie shook her head.

"Do you want to?"

She took another gulp of tea. "I don't know. It's confusing. He was a conchie…"

"Mmm. So Margaret Hamilton said." Mrs. Nelson chuckled as she stirred her tea. "She was on a tear that morning—told me it was a sin to have that conchie in Captain Nelson's bed—that I'd have to burn the sheets—that cowardice could never be washed out…"

"He wasn't a coward!"

"No?" Her tone, matter-of-fact, not accusing, made Hattie want to explain.

"He was soft and gentle and at first I thought he was weak and I hated him. Then when Mima got sick and I had so much to do, and Johnny wouldn't do a thing—just run with that Hamilton bunch—David picked up and did things when Dada didn't have him busy. Carried the water, weeded the garden—even things Will never did, like cut fruit for canning.

"Then the night Eli… David stood up to the whole gang… just to protect that little toad. He knew they might have killed him, so he couldn't be a coward, could he?" She paused as if waiting for an answer. "But why was he a conchie in the first place?"

Mrs. Nelson pushed her teacup aside. "People always told me John was courageous—especially after he was killed. At the time, I wanted to kick them in the shins. I didn't want to hear about his courage, not when he'd died and left me the way he had. To be truthful, I didn't think he was brave at all, just bone stupid."

Mrs. Nelson straightened in her chair and her eyes looked, Hattie thought, as if someone had stoked a fire behind them. Maybe her neighbor had chosen the right profession for herself. Right now, she looked like she could give a fire and brimstone sermon for Pentecost Sunday.

"I think John had a kind of courage—he and all those sad boys who signed up to fight—the courage to go all the way over there to face God knows what. They were cut down like steers in an abattoir—only they weren't steers! They were men with talents and passions and sweethearts—men who could build bridges and heal bodies and be tender to a woman in the night. They had a kind of courage and they died and we should remember them for that."

Hattie leaned forward. "Courage is for the dead, then?" she whispered. "What about the rest of us?"

She had never seen Mrs. Nelson shed a tear, but now one flickered across the older woman's eyelashes, disappearing in a quick move of her hand. "I'm sorry. I do climb up on my soapbox." She reached across the table for Hattie's hand.

"There's enough kinds of courage for all of us. Right now, we need what David may have had all along. We need the courage to put one piece of cutlery in the drawer after another until the drawer is full."

Hattie yanked at the lace on her blouse. All of a sudden the air in the kitchen was close and sweltering "I'm tired of putting everybody's cutlery away!" She swept from the table, trying to catch a breath through the screen on the door. "Sometimes I want to take the whole drawer and dump it on the floor for someone else to take care of."

Mrs. Nelson put a thin cool hand on Hattie's arm. "I'm not talking about doing for others. I'm talking about doing for yourself. You're the only one who can pack your own valise."

Something heavy crashed against the front of the house, and Hattie flew outside, circling round. Johnny—on the front porch—so out of breath Hattie could scarce make out what he said. Bent over, his hands on his knees, gulping air, he spat out a word every time he exhaled.

"Will... Da... today... station."

Mrs. Nelson rubbed his back and made him sit at the kitchen table for a cup of tea before he could tell them the whole story. Will arrived today by train; Dada had already left for the station.

Hattie grabbed her shawl and tore through the woods. The kitchen at home was in an uproar. For the first time in weeks, her mother was making bread. Laid out on the table were the sugar and cocoa she had hidden away for Will's homecoming. Her mother had killed a fat chicken, and it flopped in a corner, waiting for Hattie to pluck and clean it. Mounds of dusty potatoes and carrots crowded the small counter along with a slab of bacon from the smokehouse.

Hattie put on an apron and worked alongside her mother

until the meal was ready, the house fragrant with the smells of fresh bread, roast chicken and chocolate cake. At least the work helped the time pass. They were barely done when the buckboard wagon turned up their drive with a passenger sitting beside Dada.

Her mother's hands fluttered briefly with her apron strings, then stopped as if the task were too difficult. Hattie followed her mother onto the porch and into the yard. Johnny slammed the barn door and streaked across the yard.

Hattie held her breath, and it seemed to her as if her heart had stopped too. Johnny hooted and launched himself at Will as if he'd forgotten how much bigger he'd gotten in three years. Mima's hands trembled, and she hugged Will as if she couldn't let go. Eventually she did step back, but she continued to touch Will's jacket lightly but reverently, as if it were the relic of a saint.

Then it was Hattie's turn. He hugged her fiercely, pulling her tight against his smoky coat. But when she stood back and looked into his eyes, it was as if a stranger had come back to her. His look was sad and puzzled, as if he were searching for the answers to many questions. She touched his cheek with her hand, but the Will she had waited for had not come back after all.

Dinner was the worst. Her parents hadn't been that light-hearted since well before the war. It was as if they didn't notice how Will had changed. Mima hovered like a robin over a worm, watching every move Will made. Dada was unusually talkative, mostly about plans for the farm now that Will was home.

Johnny was incorrigible. He had to see all Will's scars: the gash on his thigh where the shrapnel had hit, the shallow ridge in his skull where the bullet had passed, and the gap in his right ear where the rat had chewed a piece off.

After the meal, her father pushed back his chair and loosened his belt. "Well, son, now that you're home, maybe your mother and I will retire."

Johnny snorted, but Dada paid no attention.

"There's a nice little house on Water Street would be just right for us," he continued. "Has a full carpentry workshop in the back, tools and all. I could make furniture and do enough handiwork to get us by. And Johnny could go to high school."

"But all my friends are out here," Johnny protested.

Dada just smiled at Will. "But we don't have to decide anything for a while yet."

Hattie slipped out to the verandah. It had not escaped her notice that her father's plans did not include her. Will would get the farm, Johnny would go to school and Hattie would, what? Go to the devil? Or maybe they figured she would continue to play nursemaid to her mother, who, despite her energy today, was still not herself.

For the first time ever, three Tamblyn men went to the barn to do chores. When they emerged sometime later, Johnny set off with the milk cans toward the river. Her father came up the stairs and into the house, whistling softly to himself. Will went straight for the woods. Hattie knew where he was going and she followed.

He was sitting on the rock waiting for her. She sat beside him. This rock had been so large when they were kids. The whole Sherwood Band had sat for picnics on this rock. Will picked up her thoughts.

"Only big enough for two of us now. Course there's not many more than that left, is there?" He pressed his finger-tips together and stared at his hands. "Tom was first, drowned in a mud hole at Passchendaele. Wouldn't have had a body to bury neither, except we knew it was him when he floated to the top with his puttees on backwards. He always did that when the officers weren't looking, wound them top to bottom like the artillery.

"Then Frankie…and Jimmy." He glanced at Hattie. "Herbie's the only one came out without a scratch. And I'd figured he was a goner the minute he signed up for Field Ambulance. Carrying wounded men on stretchers for miles across that gumbo, shells pounding all around. You never can tell."

Hattie didn't know what to say to him. She couldn't help but remember how he'd been before he left. All teasing and charm. "Don't worry, Hattie, we'll all be back to torment you soon. Even Jimmy, once the slacker gets over the measles."

Now he sat bent over, fingering a small, red leather pocket diary. "I can't do it, Hattie."

She shrugged. "Do what?"

"Stay here on the farm. I can't."

"Why not?"

He waved the diary, his voice on the edge of bitterness. "It's Jimmy's—he bought it in the canteen in Halifax—specially made for Canadian troops. Canadian facts and figures, household tips—so we wouldn't forget where we came from. He wrote in it every day the first year, little things: bath parade, pay parade, who he wrote to, who wrote to him.

"Then one stretch on the line, we didn't have anything to read. And we were in a real hole, ankle deep in water and cold as hell. Jimmy read the front section out loud to us: 'British Columbia, the Switzerland of Canada, traversed by four ranges of mountains with intervening valleys and plateaus.'" He flipped through a few pages. "'To remove oil marks on wall paper, apply paste of cold water and pipe clay, leave it on all night, brush off in the morning.'" He handed her the diary.

"You wouldn't believe how funny a thing like that sounds at three in the morning when you're up to your arse in mud. But the best part was when he'd read the names of all the towns and their populations. 'Lunenburg, Nova Scotia, 1,006. Indian Head, Saskatchewan, 1,285.'

"He was trying to decide where to go after the war—trying to figure out which place appealed to him most. We'd lay there after he'd read out a name and try to imagine what the place looked like. Did it have a hotel, or a rooming house with a good-looking widow-woman? Were there dances on Saturday night with a fiddler?

"For most of us, it passed the time, but Jimmy took it seriously. One night, just before we went back on the line, we had eggs and potatoes at an estaminet near Arras. He told me he'd decided where he was going to go. A solemn vow he took—said that if he got through the war, he was going to Great Whale River in the Northwest Territories.

"The next day, we went back to the front. He volunteered for the advance party, and I never saw him again."

Will reached for Hattie's hands, pressed them tight, the diary cold and smooth against her left hand. "I can't stay here and pretend it didn't happen. I owe it to him, don't you see?" His voice dropped to a whisper. "When Jerry put on the big push in the spring, some of the fellows… saw … lads beside them, fighting… lads that had died months before."

"Did you …" She squeezed his hand.

He shook his head. "But I feel him, like he's out there. Waiting for me to do the right thing. I have to go to Great Whale River, like he would have. I owe him much more than that. But now that he's gone, it's the only thing left I can do for him."

Will sank to the ground and put his head in her lap. She put one hand on his curly hair and let the feelings flow over her like water rushing over rocks in the river bottom. Pity for everything he'd lost, anger at everything she'd given up waiting for him to come home, fear for what would happen to her now.

May 1919
DAVID

DAVID FOUND THE HOUSE in a row of brown-stones off Roncesvalles Avenue.

"I'm looking for Mrs. Liedermann," he told the woman in the starched apron who answered the door. "My name is David Ross."

She ushered him in and bade him wait in the front hall. David paced the small entryway. He wasn't sure he should be here.

Mrs. Campbell from the church had visited him in hospital as his broken leg recovered from Thatch Hamilton. She was the one to tell him that Franz Liedermann had died of influenza and he'd felt cold inside. He couldn't even bear to think about music—every strain of melody, every bar of notes reminded him of what he'd lost. He'd wanted to see Mrs. Liedermann, knowing she was the one person who would understand how he felt. But by the time he was walking again, they said she'd gone crazy with grief. He had an idea what that meant, the way Hattie's mother had faded like a piece of fabric washed too many times. Mrs. Campbell, though, had given him the address of the rest home and told him very directly that he should pay a visit. David hoped he was doing the right thing.

It seemed a very long time before the woman returned to usher him into a side parlor of the large house. She paused at the door.

"The doctor has made it very clear she is not to be over-excited. You do understand?"

David nodded and followed her in. Mrs. Liedermann sat in one of a pair of rose wing-backed chairs by the fireplace. Always slim, she was now gaunt and pale. She reached across with both hands as he sat down and held his right hand for a long minute.

The woman in the starched apron cleared her throat. "Ada," she murmured, "don't wear yourself out." She smoothed imaginary wrinkles from her apron with both hands and nodded at David, reinforcing her warning, before she withdrew.

Mrs. Liedermann sighed softly and leaned back. "They don't understand. From the time I was first in love with him, he is part of every thing I do, every thought I have. I cannot change that."

David didn't know what to say. He traced the pattern on the arm of the chair over and over with his finger.

She smiled faintly and her eyes brightened. "It was the smallest of things that made me know I was in love with him. Do you mind an old woman's foolishness?"

"Uh-uh." He shook his head.

"His bare foot—such a thing. I told him years after and we laughed. He was always quick, going here and there, but often we would stop at this outdoor cafe. And he would line up at the bar to get our tea, talking with the bartender, the guests. Everyone loved him. I wasn't sure, you see, if I wanted to be married to someone that everyone loved. But this one day, a hot summer day. He wore these soft leather shoes but no socks. As he leaned on the bar, talking and smiling, he lifted one foot out of his shoe—a casual thing, an ordinary thing. But somehow, the curve of his tanned foot caught at my heart. And I knew I was in love with him."

David knew what she meant. The curve of Hattie's cheek the day she told him about the letters. Would that memory, he wondered, be as strong, as sweet, fifty years from now as Mrs. Liedermann's was?

"So when they took him to the camp, I worried of course, but the small things he always did stayed with me—the way he scratched his beard when he was thinking, the way he struck the table with his finger to make a point, the smile he saved for moments when he thought no one could see him. They told you I am crazy, yes?"

David shrugged. How could he tell her?

"I know what they say. She says to me lots of women have lost someone, husbands, sons. Why should I think I'm special? She doesn't understand. I have not lost Franz." She pressed her hands flat to her chest. "He is here, in every breath I take."

David lowered his eyes in respect as her breathing changed, a sharp intake of air, a quick exhale, as if she was fighting back tears.

"Every night since they told me he was dead, I have the same dream. I am sitting across from him. His hands are flat on the table—he did that when he explained something."

David nodded.

"My hands are on the table too, just a half an inch from his. I want to touch his hands, but I can't. I try for what seems like hours, but my fingers are frozen to the table and nothing will let me touch him one last time. She says I wake them in the night, but she doesn't understand. He died without me."

They sat in silence for a long time. David remembered playing the Schubert impromptu on Mrs. Nelson's piano. How the long nights of studying the music with nothing but Mr. Liedermann's teaching to guide him had resulted in sounds so powerful and sweet that they'd melted Hattie's anger, and she'd held his hand in the moonlight.

"He's in everything I play," he said finally. "He'll be in everything I'll ever play."

Mrs. Liedermann leaned forward and grazed his knee with her fingertips. "You were special to him, David. He loved to watch you learn."

Leaning to one side, she pushed a brown valise toward him. He recognized it at once. "He would want you to have this," she said.

David undid the buckle. The worn case was stuffed with sheet music. Pulling one loose, he ran the unfamiliar notes through his head. He could almost hear Mr. Liedermann, as if he stood behind him, 'You can learn it, David. Listen for the notes.'

Their eyes met. As much as this music meant to him, how much did it mean to her? "Are you sure?" he asked.

"Oh, yes. He would want it."

"Is there anything I can do?"

She put one hand on his arm. "Take the music and when some time has passed, come back and play for me. And we will remember him together."

The woman in the starched apron reappeared as he closed the parlor door behind him.

"Take good care of her," he said, trying for the first time to copy his uncle's tone of command.

The woman looked thoughtful. "We will."

She followed him onto the porch and as he descended the stairs, she added: "There are a great many women, I think, would be grateful for the gift of a dead husband. The perverseness of the world, I suppose, that hers should be taken instead of one of them."

David took care not to swing the valise as he headed toward Roncesvalles, wondering if the woman in the starched apron was married and what Hattie would think if he wrote to her.

June 1919

HATTIE

*D*ADA GAVE HER THE LETTER as she finished the milking.

"This came yesterday, guess I forgot."

She looked at the postmark. Toronto.

"Can't be from Will," she said as he waited for her to open the letter. "He'll be in Calgary or Edmonton by now." She slipped the fine linen envelope in her apron as if the mystery didn't tempt her at all.

Who was she fooling? She knew who it was from; she was just afraid of what it might say. In his last letter, he'd asked her, "What do you want, Hattie Tamblyn?"

Sitting at the small desk in her room late at night, her answer had poured all over the page. She'd mailed it before stopping to consider what he'd think. Was she too fierce? Too passionate? She could remember the words as if they were carved in her brain, always there for her to touch.

"When I was little, I wanted to be Little John and knock the blocks off the Sheriff's men. But hitting Eli made me sick after, especially when I saw what Thatch had done to him—when he came groveling, giving back your fifty dollars.

"When Jimmy died, I was the only one left to be Little John. But I never wanted to be Little John by default.

I wanted to be Little John because I was the best. But what am I the best at now?

"I thought I wanted what Will had—the independence to get up and go away, but going away scares me now because I have nothing to go to. I think about how you love your music and how you worked so hard at it when you didn't even have a piano. And I know that's what I want—something of my own that I can be best at."

Late in the long evening, as the sun was smearing its current with gold, Hattie went down to the river. Slipping the letter from the envelope she read it again and again. Not an answer, but a way to find the answer.

"Mrs. Campbell," David wrote, "says such decisions are best made without undue pressure. She says you can stay until you're sure."

Folding the envelope into a boat—a coracle like the old Irish monks used to sail to Iceland—Hattie slipped the dice inside and plied it into the current. She wanted them to be far away by the time she left for Toronto. She watched the tiny coracle disappear around the curve in the river.

In her last letter, she'd told David she wasn't brave enough to set off on her own. Tonight, she'd read his reply. "I don't think courage is something we have. I think it's something we share. I don't think I would have believed in myself if Mr. Liedermann hadn't believed in me first."